Electric Motors and Control Systems, 3e

Activities and Simulations Manual

Frank D. Petruzella

ELECTRIC MOTORS AND CONTROL SYSTEMS, THIRD EDITION
ACTIVITIES AND SIMULATIONS MANUAL

Published by McGraw-Hill Education, 2 Penn Plaza, New York, NY 10121.

Some ancillaries, including electronic and print components, may not be available to customers outside the United States.

This book is printed on acid-free paper.

4 5 6 7 8 9 LKV 24 23 22

ISBN 978-1-260-43938-0
MHID 1-260-43938-0

mheducation.com/highered

Contents

Preface

This manual is divided into two distinctive parts as follows:

Part 1—ACTIVITIES MANUAL contains hundreds of objective type quizzes and practical assignments for each chapter in the text.

Part 2—THE CONSTRUCTOR simulation motor control program. This special edition of The Constructor program contains preconstructed simulated motor control circuit assignments. All assignments are reference to the associated circuit diagrams found in the text.

ANALYSIS type assignments are designed to enhance your understanding of the motor control circuits, the operation of which, are covered in the text.

TROUBLESHOOTING type assignments provide an opportunity to apply your knowledge of the motor control circuit, by use of a test probe to identify faulted components.

The Constructor motor control simulation software can be downloaded from CMH Software Inc., using the link and access code available with this text. (Note: Those using an electronic copy of this text can access the software at http://highered.mheducation.com:80/sites/0073373818.) This download of the program contains the pre-constructed simulated motor control circuits in both NEMA and IEC format.

I would like to thank the team from CMH Software Inc.—Bob Hosea and Kevin Christensen—for their collaborative effort in the development of the simulated motor control circuit analysis and troubleshooting assignments. I would like to thank McGraw Hill Product Developer Tina Bower for overseeing the development of this manual, and Don Pelster from Nashville State Community College for an outstanding job of editing the manual.

Frank D. Petruzella

Part 1

ACTIVITIES

1 Safety in the Workplace

PART 1 Quiz: Protecting against Electric Shock

Place the answers in the space provided.

1. The higher the body resistance, the greater the potential electric shock hazard. (True/False) F
2. Moisture Decrease (increases/decreases) resistance.
3. Generally any voltage above 30 V is considered dangerous.
4. An electric shock with a current of 100 mA is rarely fatal. (True/False) F
5. The pathway taken by electricity through the body has little influence on the severity of an electric shock. (True/False) F
6. An electric shock can result any time a body becomes part of an electric circuit.
7. The most common electricity-related injury is a burn. (True/False) T
8. The three electrical factors involved in an electric shock are resistance, amps, and volts.
9. Which of the following statements about electric shock is *not* true?
   a. The lower the body resistance, the greater the potential shock hazard.
   b. The higher the voltage, the greater the potential shock hazard.
   c. The lower the current, the greater the potential shock hazard.
   d. The longer the length of time of exposure, the greater the potential shock hazard.

   Answer C
10. The amount of current flowing through a body is equal to
    a. E/R
    b. $E \times R$
    c. P/R
    d. $P \times R$

    Answer A

11. ______________ percent of electrical workplace accidents are caused by arc flash.

    a. 80
    b. 50
    c. 25
    d. 5

    Answer A

12. Which of the following is not a common cause of arc flash?

    a. Build-up of conductive dust and dirt
    b. Inadequate personal protection equipment
    c. A dropped tool
    d. Accidental contact with electrical systems

    Answer B

13. Activities with high potential for exposure to arc flash include

    a. Troubleshooting live circuits
    b. Opening or closing circuits
    c. Grounding circuits
    d. All of these

    Answer D

14. The primary method for preventing employees from arc flash is to deenergize live parts prior to working on them using proper lockout/tagout procedures. (True/False) T
15. You should avoid wearing polyester (polyester/cotton) clothing on the job.
16. Leather gloves are worn over rubber electrical gloves to protect the rubber glove from punctures. (True/False) True
17. Hard hats are made of a nonconductive material as an added protection against electric shock. (True/False) True
18. Receiving an electric shock is a clear warning that proper safety measures have not been followed. (True/False) T
19. Keep material or equipment at least 10 feet away from high-voltage overhead power lines.
20. When installing new machinery, ensure that all metal framework is efficiently and permanently grounded.
21. Capacitors should be discharged before handling them.

22. A safety light curtain is designed to provide workers with acceptable lighting required to operate machinery. F
23. A normally closed general-purpose push button switch is not an acceptable emergency stop switch. T
24. Which of the following safety solution is most suitable for safeguarding the interior of robotic work cell?
    a. safety laser scanner
    b. emergency stop controls
    c. safety interlock switches
    d. safety light curtain
25. Which of the following safety solution is most suitable for monitoring the position of a guard or gate?
    a. safety laser scanner
    b. emergency stop controls
    c. safety interlock switches
    d. safety light curtain
26. What must be performed immediately after locking out electrical equipment?
    a. remove fuses
    b. documentation actions
    c. perform an infra-red scan
    d. check zero energy
27. Safety light Curtain create a curtain of photoelectric light beams between an emitter and a receiver.
28. The self-holding function of an emergency stop requires that the control process cannot be started again until the actuating stop switch has been Switched to the on position.

PART 2 Quiz: Grounding—Lockout—Codes

Place the answers in the space provided.

1. Grounding is the intentional connection of a current-carrying conductor to the earth.
2. Bonding is the permanent joining together of all non-current-carrying conductive enclosures.
3. The purpose of grounding is to
    a. Limit voltage surges
    b. Provide a ground reference that stabilizes
    c. Facilitate the operation of overcurrent devices
    d. All of these

   Answer D

4. Bonding establishes a low-resistance path for ground fault current. (True/False) T

5. Bonding limits the touch voltage when non-current-carrying metal parts are inadvertently energized. (True/False) T

6. The resistance of the earth as an electrical conductor is very low. (True/False) F

7. The two distinct parts of a grounding system are the System grounding and the equipment grounding.

8. System grounding is the connection of one of the current-carrying conductors to ground. (True/False) T

9. Which conductors form part of the grounding system?

   a. Equipment grounding conductor
   b. Grounded conductor
   c. Grounding electrode conductor
   d. All of these

   Answer D

10. A ground fault is an unintentional connection between an ungrounded conductor of an electric circuit and metallic enclosures and raceways.

11. A ground fault circuit interrupter (GFCI) is a device that senses small currents to ground and acts to quickly shut off the current.

12. A GFCI is capable of reacting to leakage current as small as 5 mA.

13. A GFCI protects against

   a. Line-to-line short circuit
   b. Ground fault
   c. Overload
   d. All of these

   Answer B

14. A GFCI monitors

   a. The current flow in the live conductor
   b. The current flow in the neutral conductor
   c. Any difference in current flow between the live and neutral conductors
   d. The sum of the current flow in the live and neutral conductor

   Answer C

15. Electrical lockout refers to the process of _______________ the power source in the off position.
16. Electrical _______________ is the process of placing a danger tag on the source of electrical power, which indicates that the equipment may not be operated until the danger tag is removed.
17. Electrical lockout/tagout is at times used in servicing electrical equipment that requires the power to be on. (True/False) F
18. "Dead" circuits should be
    a. Checked by momentarily touching them
    b. Checked by metering the voltage
    c. Locked out and tagged out before starting any repairs
    d. Both b and c

    Answer D
19. A lockout/tagout must not be removed by any person other than the authorized person who installed it. (True/False) T
20. A disconnect switch should not be operated if the switch is still under load. (True/False) T
21. There should be only one lock and tag on the disconnect switch even if more than one person is working on the machinery. (True/False) F
22. OSHA, the Occupational Safety and Health Administration, is a federal agency whose mission is to ensure the safety and health concerns of all American workers. (True/False) T
23. The National Electrical Code (NEC)
    a. Outlines the minimum requirements that must be met for a safe wiring installation
    b. Outlines the ideal requirements for a wiring installation
    c. Is a manual designed to teach how to properly wire houses
    d. Both a and c

    Answer A
24. Article _______________ of the NEC deals with motors and all associated branch circuits.
25. Hazardous locations include petroleum refineries and gasoline storage and dispensing areas.

26. Use a class ______________ –rated fire extinguisher on energized electrical wiring or equipment.

  a. A
  b. B
  c. C
  d. D

  Answer C

27. In case of fire in the workplace, it is your responsibility to

  a. Know the safest means of exit from the building
  b. Leave the danger area in an orderly manner
  c. Trigger the nearest fire alarm
  d. All of these

  Answer D

28. ______________ or ______________ indicates that a piece of electrical equipment or material has been tested and evaluated for the purpose for which it is intended to be used.

29. Which of the following is a recognized testing laboratory?

  a. UL
  b. NFPA
  c. NEC
  d. All of these

  Answer A

30. Which of the following develops standard frame sizes for motors?

  a. NEMA
  b. IEC
  c. IEEE
  d. Both a and b

  Answer D

2 Understanding Electrical Drawings

PART 1 Quiz: Symbols—Abbreviations—Ladder Diagrams

Place the answers in the space provided.

1. In a ladder diagram, components are arranged according to their physical position in the installation. (True/False) false
2. When conductors cross but make no electrical contact, this is represented by intersecting lines with no dot. (True/False) true
3. Circuits are normally represented on diagrams in their deenergized position. (True/False) true
4. The conductors feeding current to a large motor would be considered to be part of the power circuit. (True/False) ~~false~~ True
5. When more than one electrical load is connected in a rung of a ladder diagram, the loads are normally connected in series. (True/False) false
6. All additional stop or off control devices must be wired in parallel. (True/False) ~~true~~ False
7. Identify each of the symbols shown in Figure 2-1:

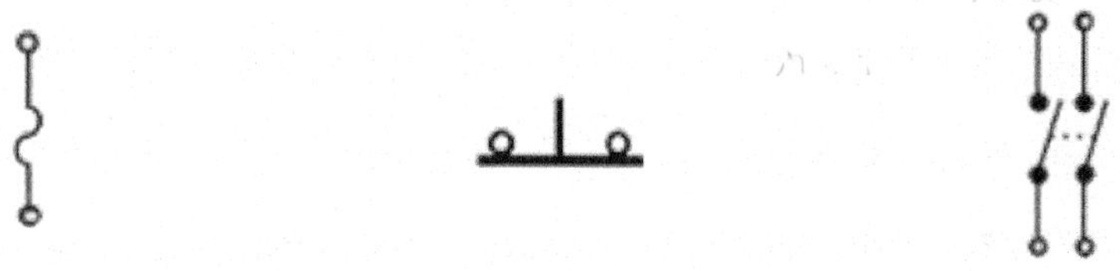

a. ~~OL relay~~ Fuse b. Push button normally closed c. 2 pole non fused switch

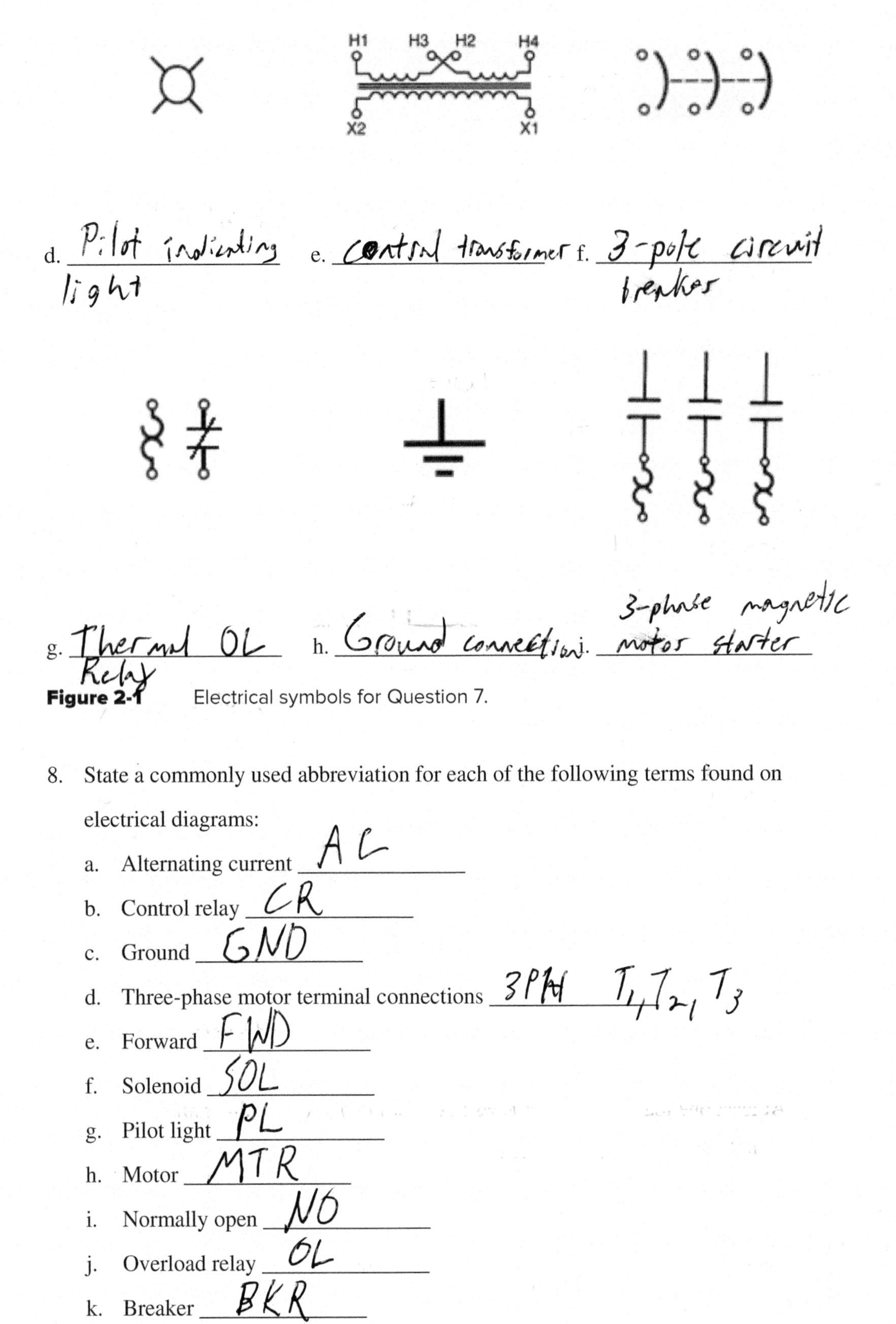

d. Pilot indicating light e. control transformer f. 3-pole circuit breaker

g. Thermal OL Relay h. Ground connection j. 3-phase magnetic motor starter

Figure 2-1 Electrical symbols for Question 7.

8. State a commonly used abbreviation for each of the following terms found on electrical diagrams:

   a. Alternating current AC
   b. Control relay CR
   c. Ground GND
   d. Three-phase motor terminal connections 3PH T_1, T_2, T_3
   e. Forward FWD
   f. Solenoid SOL
   g. Pilot light PL
   h. Motor MTR
   i. Normally open NO
   j. Overload relay OL
   k. Breaker BKR

l. Rectifier Rec

m. Positive Pos

n. Push button PB

o. Rheostat RH

p. Secondary SEC

q. Transformer trans

r. Time delay TD

s. Phase PH

t. Manual Man

9. On the ladder diagram shown in Figure 2-2, complete the identification of the rungs and the wire numbers.

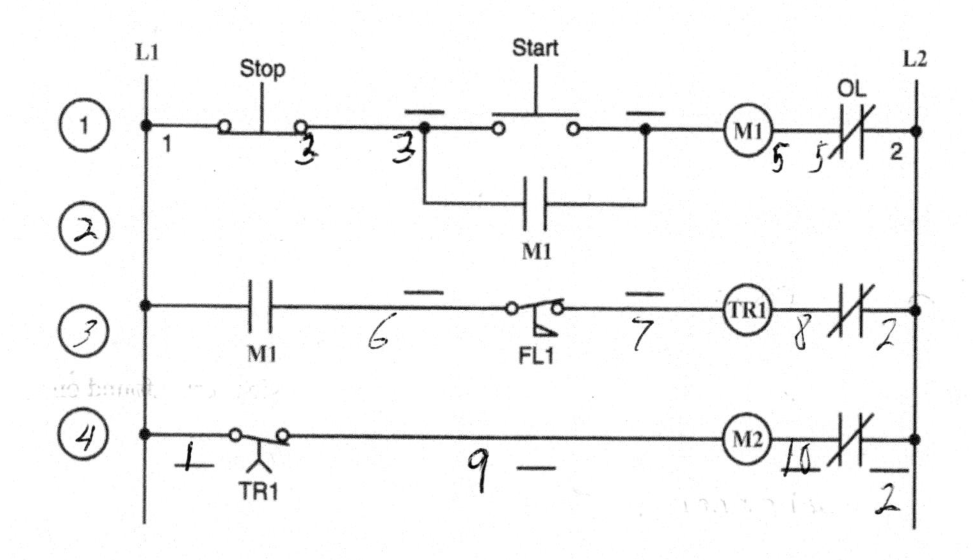

Figure 2-2 Ladder diagram for Question 9.

10. The load is the electrical device in a ladder diagram that uses the power supplied from L1 to L2.

11. At least one load must be included in each rung of a ladder diagram.
(True/False) true

12. A broken connection on a ladder diagram indicates a ~~open~~ mechanical function.

13. NEMA and IEC motor control diagrams use different types of symbols to represent components of a motor control system.
(True/False) ~~False~~ ~~True~~ False

PART 2 Quiz: Wiring—Single-Line Diagrams—Block Diagrams

1. A wiring diagram is intended to show the actual __location__ of all components.
2. Wiring diagrams are used in installing and tracing wires in electrical installations. (True/False) __true__
3. Frequently, there will be a wiring diagram inside a magnetic motor starter enclosure. (True/False) __true__
4. Each block of a block diagram represents electrical circuits that perform a specific __function__.
5. The single-line diagram uses blocks along with a single line to show all the major components of a system. (True/False) ~~true~~ __False__
6. A conduit layout diagram normally includes the number and size of wires to be run in the conduit. (True/False) __true__
7. Which of the following diagrams shows an electrical installation reduced to its simplest form?
   a. Wiring diagram
   b. Ladder diagram
   (c.) Single-line diagram
   d. Pictorial diagram

   Answer __C__
8. On a block diagram, the arrows connecting the blocks indicate
   Answer __direction of current__
9. Answer each of the following with reference to the motor diagram shown in Figure 2-3 (p. 13):
   a. The two types of motor diagrams shown are a __wiring__ diagram and a __ladder__ diagram.
   b. Assuming the incoming line voltage is three-phase, 230 V, both the M coil and the pilot light would be rated for __230__ V.
   c. The auxiliary M contact will be closed whenever the motor is running. (True/False) __true__
   d. Assume the overload contact has tripped open. Pressing the start button closed will turn the pilot light on but the starter coil will not energize. (True/False) __true__

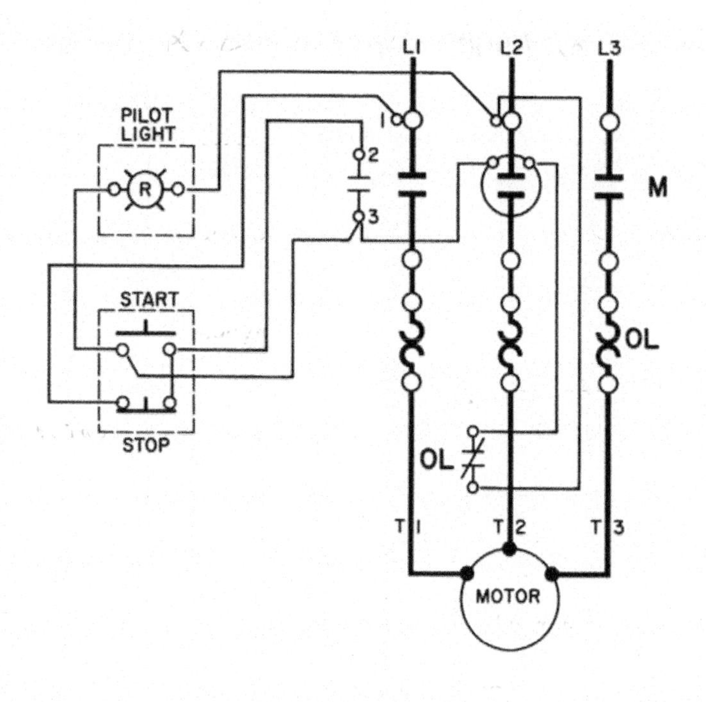

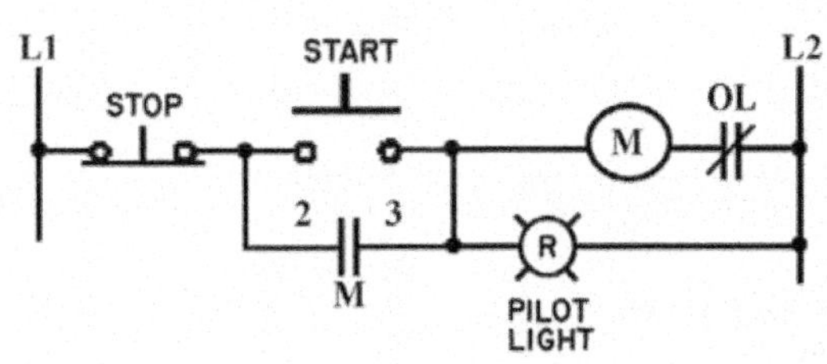

Figure 2-3 Motor diagram for Question 9.

10. Riser diagram shows the circuit as a (an) ______ diagram.
    a. elevation
    b. pictorial
    c. schematic
    d. specification

PART 3 Quiz: Motor Terminal Connections

1. Motors are classified according to the type of power used as being ~~Single phase~~ or ~~3 phase~~. AC or DC
2. For the DC motor connections shown in Figure 2-4, identify the motor type and motor leads.

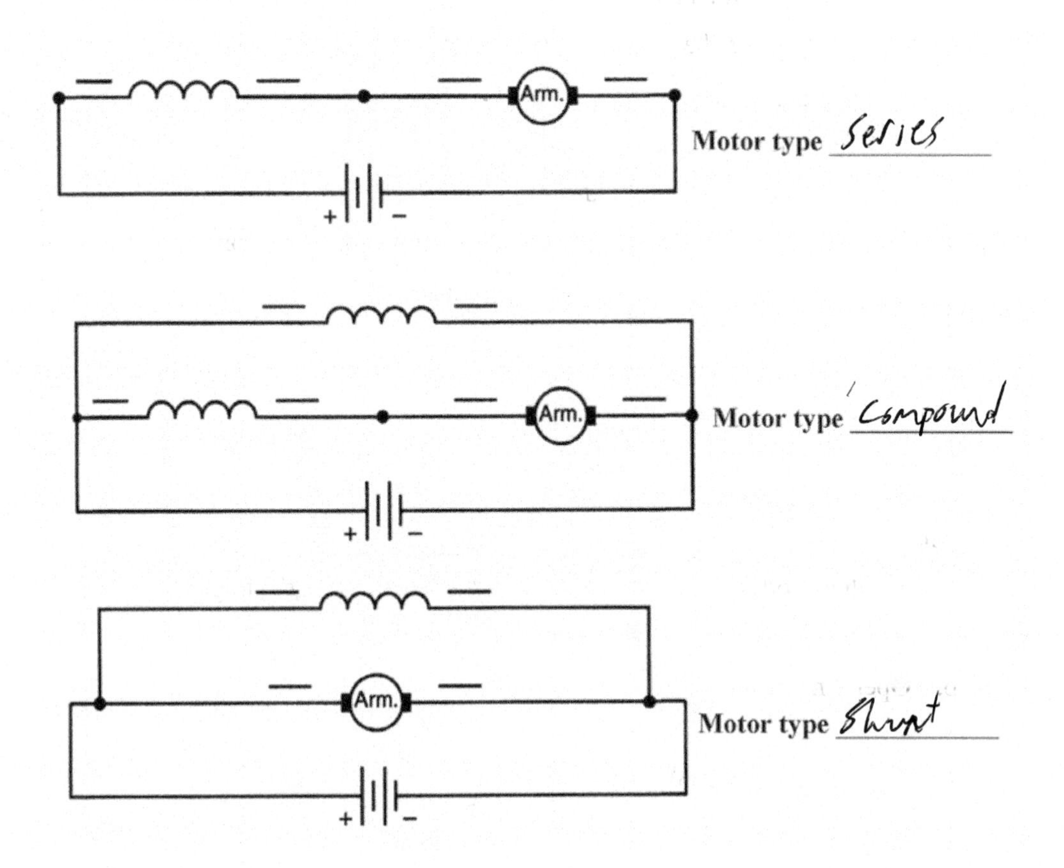

Figure 2-4 DC motor diagrams for Question 2.

3. Industrial applications use DC motors because the Speed / torque relationship can be easily varied.
4. Dynamic braking can be used with DC motors to bring the motor to a stop without the aid of a mechanical brake. (True/False) true
5. The direction of rotation of a DC motor depends on the direction of the magnetic field and has nothing to do with the direction of current flow through the armature. (True/False) true

6. The use of DC motors is more widespread than the use of AC motors. (True/False) false

7. Motor starter terminal markings are used to tag terminals to which connections can be made from external circuits. (True/False) true

8. The rotating part of an AC motor is referred to as the rotor; the stationary part is called the stator.

9. A squirrel-cage rotor has no physical electrical connection to it. (True/False) true

10. The device used to automatically disconnect the starting winding of a single-phase split-phase motor is called a centrifugal switch.

11. In a dual-voltage split-phase motor, the two run windings are connected in series for the high-voltage operation.
    a. Wye
    b. Delta
    c. Parallel
    d. Series
    Answer d

12. When a dual-voltage motor is operated at the lower voltage rating, the
    a. Operating power is lower
    b. Operating power is higher
    c. Operating current is lower
    d. Operating current is higher
    Answer d

13. When a split-phase motor is operated with a capacitor in series with one of the stator windings, the
    a. Starting torque is lower
    b. Starting torque is higher
    c. Starting voltage is lower
    d. Starting voltage is higher
    Answer b

14. Single-phase AC motors are typically constructed in large horsepower sizes. (True/False) ~~true~~ False

15. Three-phase motors are not self-starting on their running windings. (True/False) ~~true~~ False

16. The three distinct windings of a three-phase motor are referred to as

   a. L1, L2, L3

   b. Phase A, phase B, phase C

   c. Phase 1, phase 2, phase 3

   d. T1, T2, T3

   Answer ________

17. All three-phase motors are wired so that the phases are connected in either ______________ or ______________ configuration.

18. On the nine-lead, three-phase, dual-voltage motor connection diagram shown in Figure 2-5:

   a. Mark each motor lead (T1, T2, etc.).

   b. Identify each phase coil (A, B, C).

   c. Show the motor properly connected for high-voltage operation.

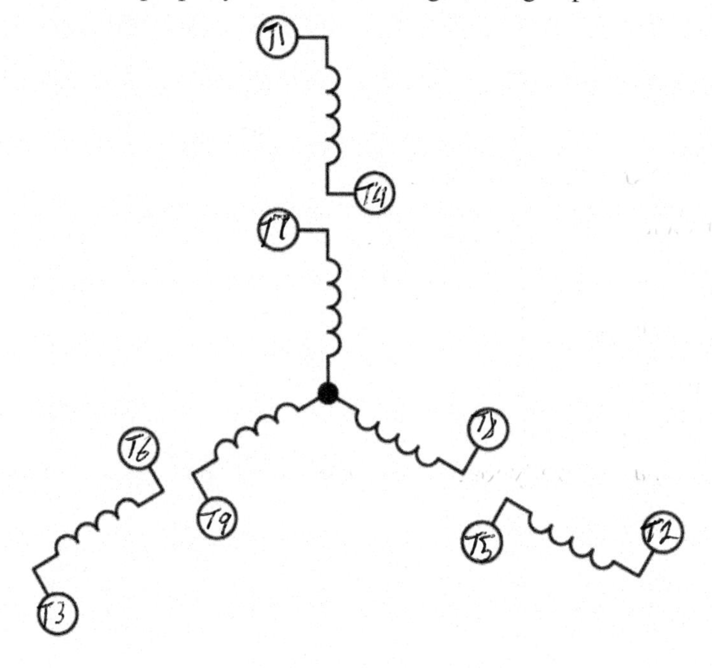

Figure 2-5 Dual-voltage motor diagram for Question 18.

19. The speed of an AC induction motor depends on the

   a. Frequency of the electrical power supply

   b. Voltage of the electrical power supply

   c. Number of poles built into the motor

   d. Both a and c

   Answer ________

20. You can reverse the direction of rotation of a three-phase motor by reversing any two of the three main power leads to the motor. (True/False) true

21. In a variable-frequency motor drive, the higher the frequency of the power supplied to the motor, the slower is its operating rpm. (True/False) ~~true~~ False

22. Inverter-duty describes a class of motors that are capable of operation
    a. From a variable-frequency drive
    b. From a DC power supply
    c. For intermittent duty only
    d. For continuous duty

    Answer A

PART 4 Quiz: Motor Nameplate and Terminology

1. Information required to identify a motor is contained on the motor nameplate.

2. NEMA requires that a motor be able to carry its rated horsepower at nameplate voltage plus or minus
    a. 2 percent
    b. 5 percent
    c. 8 percent
    d. 10 percent

    Answer d

3. Motors that are not fully loaded draw less current than their rated nameplate current. (True/False) ~~False~~ True

4. The line frequency of a motor is abbreviated on the nameplate as
    a. CY
    b. CYC
    c. Hz
    d. All of these

    Answer D

5. The phase rating of a motor is listed on the nameplate as Single phase, 3 phase AC, or Single phase AC.

6. The rated nameplate speed of a motor is its approximate speed with no load applied. (True/False) false

7. The motor nameplate ambient temperature rating is the maximum rated temperature of the air surrounding the motor.

8. The amount the motor winding temperature will increase above the ambient temperature at full load is specified according to the nameplate temperature rise rating.

9. Thermal images of electric motors reveal their
   a. Internal conductor temperature
   b. Exterior surface temperature
   c. Operating current
   d. All of these

   Answer b

10. Standard NEMA motor insulation classes are given by
    a. Number classification (1, 2, 3, etc.)
    b. Alphabetic classification (A, B, F, H, etc.)
    c. Color classification (red, blue, green, etc.)
    d. Temperature level (high, medium, low, etc.)

    Answer b

11. The duty cycle of a motor is listed on the nameplate as continuous or intermitted

12. The horsepower nameplate rating of a motor is a measure of the full-load output power the shaft of the motor can produce without stalling. (True/False) false

13. The motor nameplate code letter specifies the locked rotor-current rating of the motor.

14. The motor nameplate design letter is an indication of the motor's torque-speed curve. (True/False) true

15. A motor nameplate indicates that the motor has a service factor of 1.25. This means that the
    a. Normal routine service time interval can be increased by 125 percent
    b. Cost of operating the motor is 125 percent higher than an equivalent motor rated for a service factor of 1
    c. Motor can on occasion safely develop 125 percent of its rated horsepower
    d. Motor can safely be operated at 125 percent of its rated voltage

    Answer A

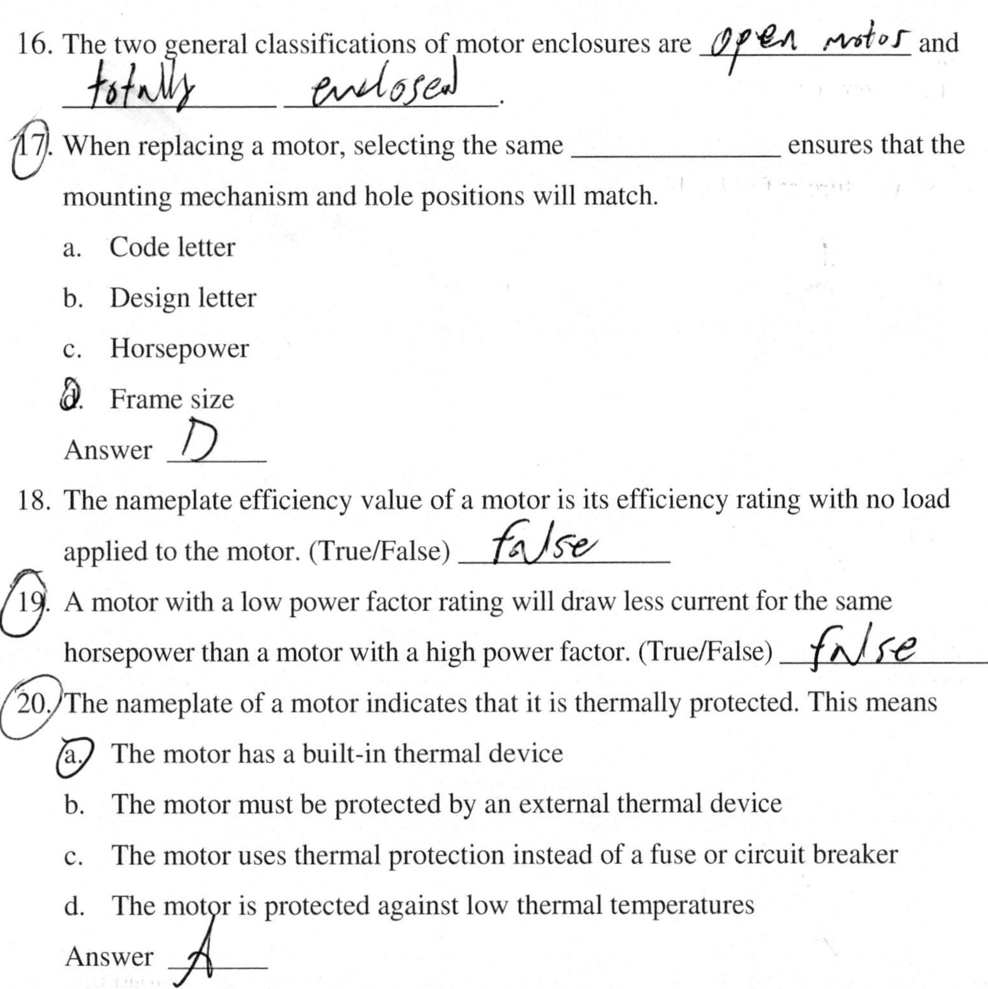

16. The two general classifications of motor enclosures are ________ and ________ ________.

17. When replacing a motor, selecting the same ______________ ensures that the mounting mechanism and hole positions will match.

    a. Code letter

    b. Design letter

    c. Horsepower

    d. Frame size

    Answer ______

18. The nameplate efficiency value of a motor is its efficiency rating with no load applied to the motor. (True/False) ______________

19. A motor with a low power factor rating will draw less current for the same horsepower than a motor with a high power factor. (True/False) ______________

20. The nameplate of a motor indicates that it is thermally protected. This means

    a. The motor has a built-in thermal device

    b. The motor must be protected by an external thermal device

    c. The motor uses thermal protection instead of a fuse or circuit breaker

    d. The motor is protected against low thermal temperatures

    Answer ______

21. Connect the motor shown in Figure 2-6 for clockwise rotation and low voltage.

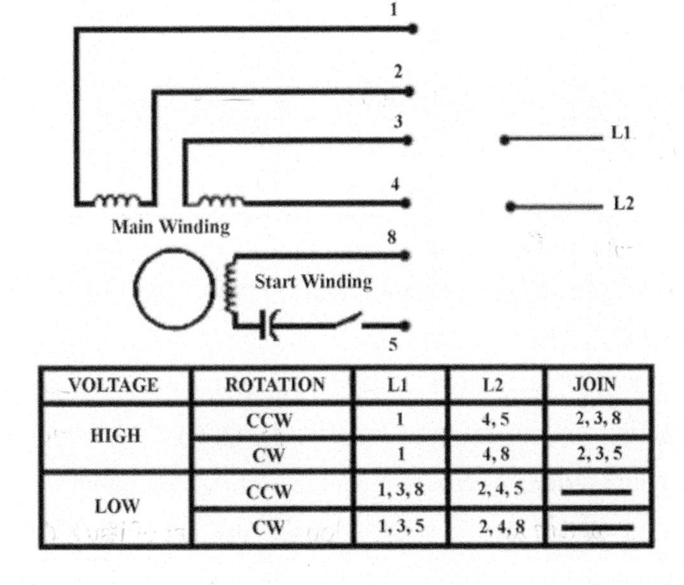

VOLTAGE	ROTATION	L1	L2	JOIN
HIGH	CCW	1	4, 5	2, 3, 8
	CW	1	4, 8	2, 3, 5
LOW	CCW	1, 3, 8	2, 4, 5	———
	CW	1, 3, 5	2, 4, 8	———

Figure 2-6 Motor connection diagram for Question 21.

22. The *plugging* of a motor refers to

    a. Braking by reverse rotation
    b. Momentary operation or small movement of a machine
    c. Automatic restarting of a motor after a power failure
    d. Applying a reduced supply of voltage to a motor during starting

    Answer A

23. The motor term *auxiliary contact* refers to

    a. A contact that is operated electromechanically
    b. A contact that is operated from some remote point
    c. The contact of a switching device in addition to the main circuit contact
    d. A contact that provides running overcurrent protection

    Answer C

24. With *low-voltage protection* motor control

    a. A power failure disconnects service and, when power is restored, manual restarting is required
    b. A power failure disconnects service and, when power is restored, the controller automatically restarts
    c. Only two wires are required for the control device
    d. Both b and c

    Answer A

25. A *motor starter* is an electric controller used to _______________ a connected motor

    a. Start
    b. Stop
    c. Protect
    d. All of these

    Answer D

26. Nine-lead dual 3-phase motors have a 2-to-1 voltage ratio.

    (True/ False) true

27. For a six-lead dual voltage with a low voltage rating of 220 volts, the high voltage rating would be

    a. 277 volts
    b. 460 volts
    c. 440 volts
    d. 380 volts

28. For six-lead dual voltage 3-phase motors the 3-phase configuration used for the low voltage is delta and for high voltage is wye.

29. The IEC nomenclature for the output leads of a 3-phase motor are identified as U, V, W

PART 5 Quiz: Manual and Magnetic Motor Starters

1. Which of the following types of motor starters connects the motor directly to the supply line on starting?
   a. Start/stop starter
   b. Jogging starter
   c. Full-voltage starter
   d. Reduced-voltage starter

   Answer ______
2. With a selector switch, a
   a. Plunger actuates the contacts
   b. Button actuates the contacts
   c. Relay actuates the contacts
   d. Rotating handle actuates the contacts

   Answer ______
3. Manual motor starters allow a motor to be controlled from any location. (True/False) ______________

Answer Questions 4 to 8 with reference to the motor control circuit shown in Figure 2-7.

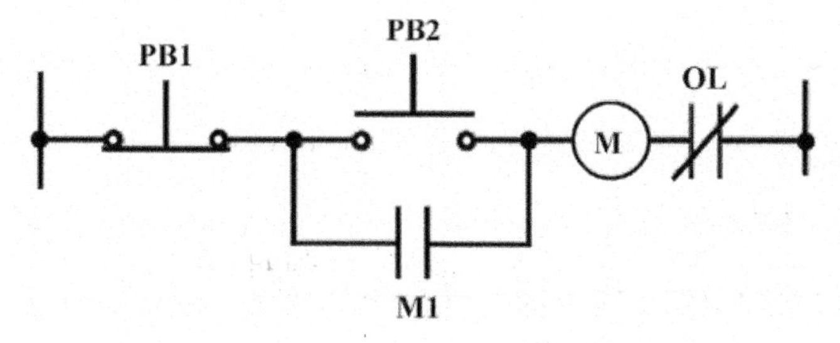

Figure 2-7 Motor control diagram for Questions 4–8.

4. Which is the normally closed push button?
   a. Start button
   b. Jog button
   c. Stop button
   d. Plug button

   Answer ______

5. Which is the normally open contact?

    (a) The holding contact

    b. The overload contact

    c. The start contact

    d. The stop contact

    Answer a

6. Where would you connect another start button?

    a. In series with the stop button

    b. In series with the start button

    c. In parallel with the stop button

    (d) In parallel with the start button

    Answer D

7. Where would you connect another stop button?

    (a) In series with the stop button

    b. In series with the start button

    c. In parallel with the stop button

    d. In parallel with the start button

    Answer A

8. This control circuit would be classified as:

    a. DC control

    b. AC control

    c. Two-wire control

    (d.) Three-wire control

    Answer D

9. Two-wire control circuits use a momentary-contact type of control device.

10. Three-wire control circuits are designed to protect against automatic starting when power returns after a power failure.

11. In the circuit in Figure 2-8, pressing the push button will:

a. Switch both lights on

b. Switch both lights off

c. Switch light A on and light B off

d. Switch light A off and light B on

Answer ___c___

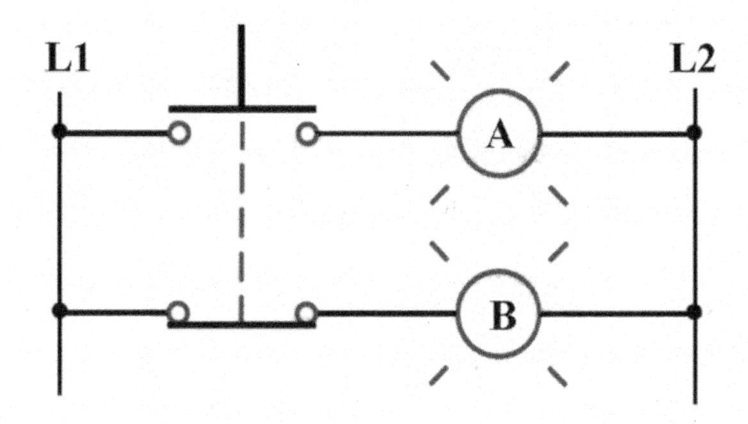

Figure 2-8 Control circuit for Question 11.

12. In the circuit in Figure 2-9, the light is switched on:

a. Only when PB1 is pressed

b. Only when PB2 is pressed

c. When either PB1 or PB2 is pressed

d. When both PB1 and PB2 are pressed

Answer ___c___

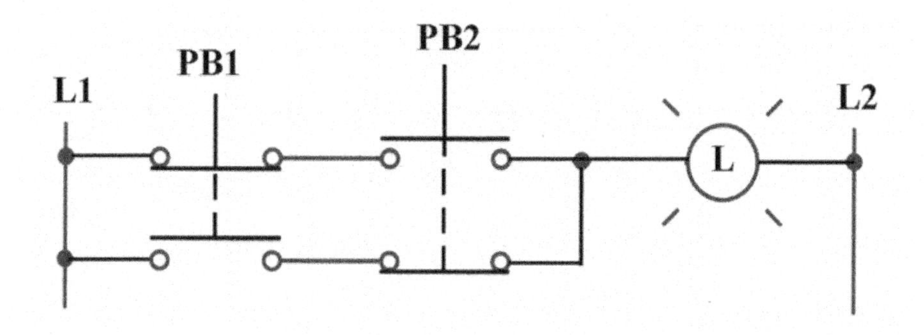

Figure 2-9 Control circuit for Question 12.

13. In the circuit in Figure 2-10, which circuit state will switch the light on?

    a. PB1 and PB2 pressed—PB3 not pressed
    b. PB1 and PB3 pressed—PB2 not pressed
    c. PB2 and PB3 pressed—PB1 not pressed
    d. PB2 pressed—PB1 and PB3 not pressed

    Answer D

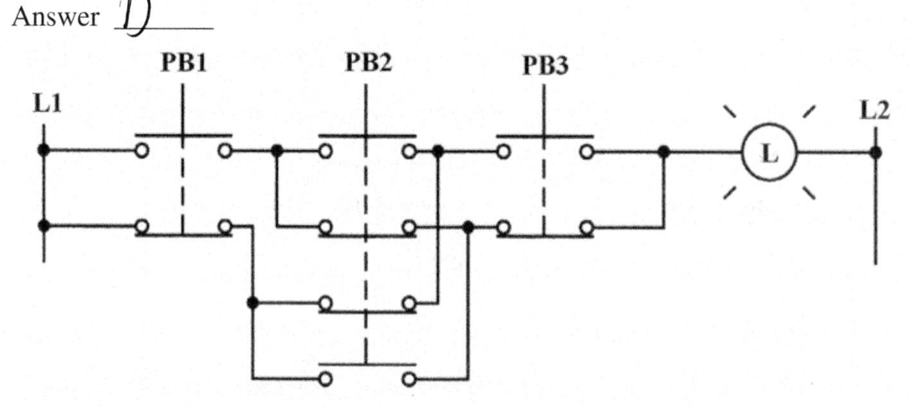

Figure 2-10 Control circuit for Question 13.

14. The operation of a 3-wire NEMA type control circuit is basically the same as that of an IEC 3-wire type control circuit.

    (True/False) True

hands on PRACTICAL ASSIGNMENTS

1. The purpose of this assignment is to develop a ladder diagram from a wiring diagram. For the magnetic motor starter shown in Figure 2-11, redraw the control circuit in the form of a ladder diagram.

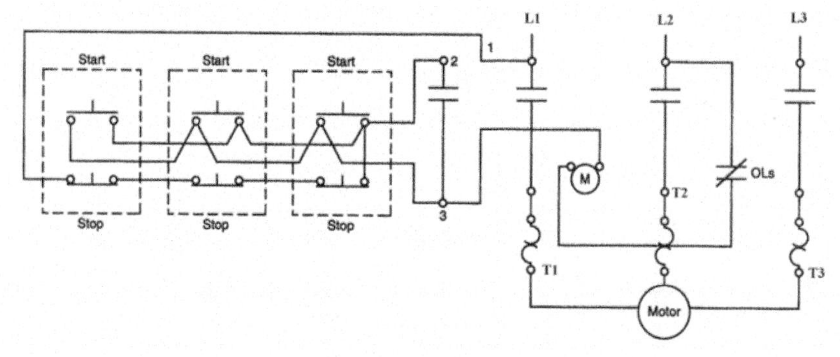

Figure 2-11 Magnetic motor starter for Practical Assignment 1.

Answer ____________

2. The purpose of this assignment is to properly connect and perform measurements on a single-phase, dual-voltage motor installation. Complete the following tasks using whatever single-phase, dual-voltage motor is available to you:
   - Connect the motor for a given voltage.
   - Reverse the direction of rotation of the motor.
   - Measure the motor current.
   - Measure the motor speed.
3. The purpose of this assignment is to properly connect and perform measurements on a three-phase, dual-voltage motor installation. Complete the following tasks using whatever three-phase, dual-voltage motor is available to you:
   - Connect the motor for a given voltage.
   - Reverse the direction of rotation of the motor.
   - Measure the motor current.
   - Measure the motor speed.
4. The purpose of this assignment is to connect a three-phase manual motor starter.
   - Complete a wiring of the circuit shown in Figure 2-12.
   - Wire the circuit in a neat and professional manner.

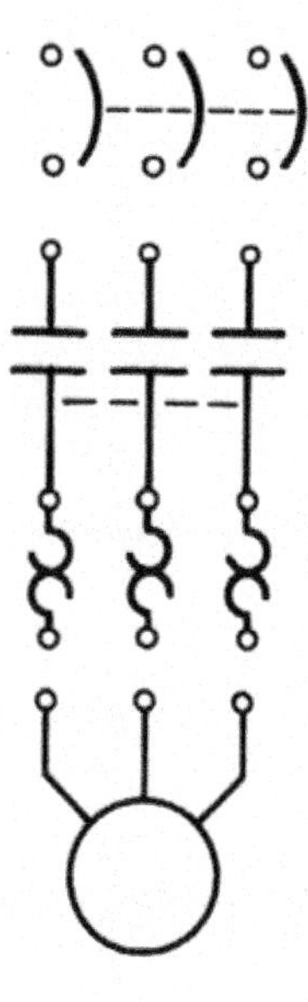

Figure 2-12 Manual motor starter for Practical Assignment 4.

5. a. The purpose of this assignment is to connect a magnetic across-the-line motor starter to operate a three-phase motor using a pilot device connected for two-wire control. Complete the wiring diagram for the two-wire control circuit shown in Figure 2-13. Wire the circuit in a neat and professional manner.

   b. Does the circuit provide low-voltage release or low-voltage protection?

   ____

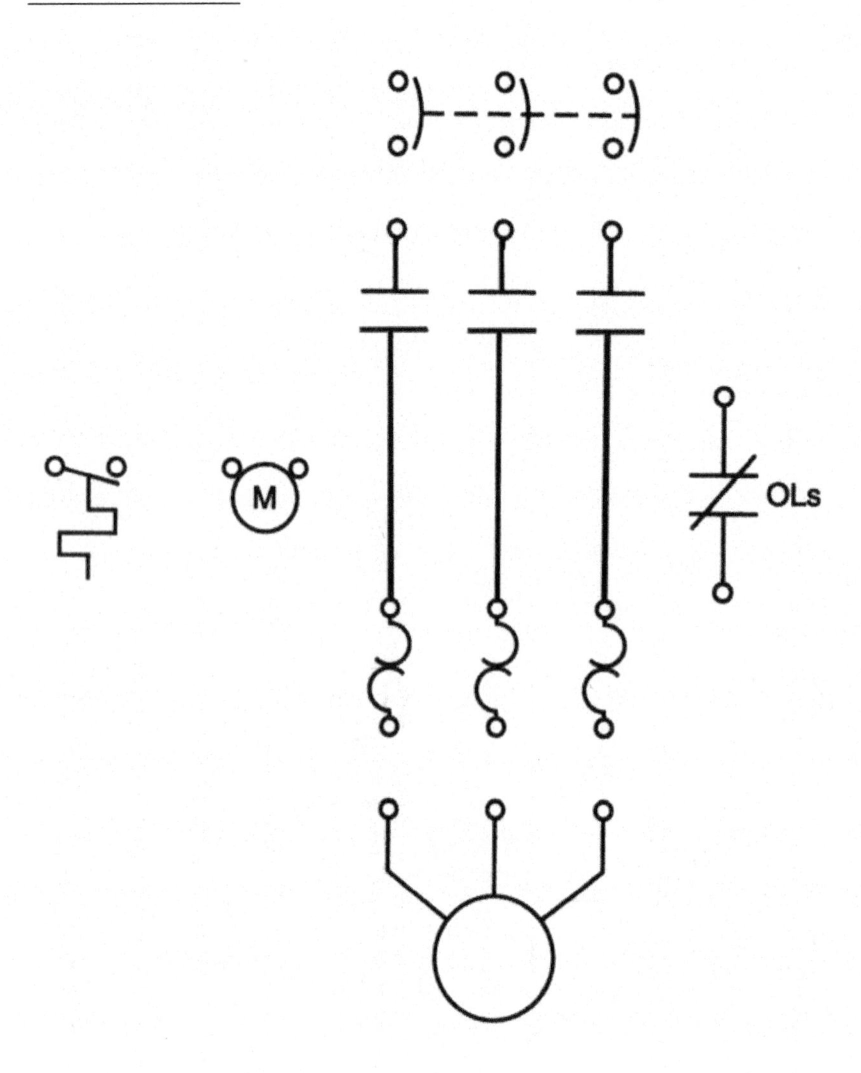

Figure 2-13 Two-wire control circuit for Practical Assignment 5.

6. a. The purpose of this assignment is to connect a magnetic across-the-line motor starter to operate a three-phase motor using a start and a stop push button connected for three-wire control. Complete the wiring diagram for the three-wire control circuit shown in Figure 2-14. Wire the circuit in a neat and professional manner.

   b. Does the circuit provide low-voltage release or low-voltage protection? _______________

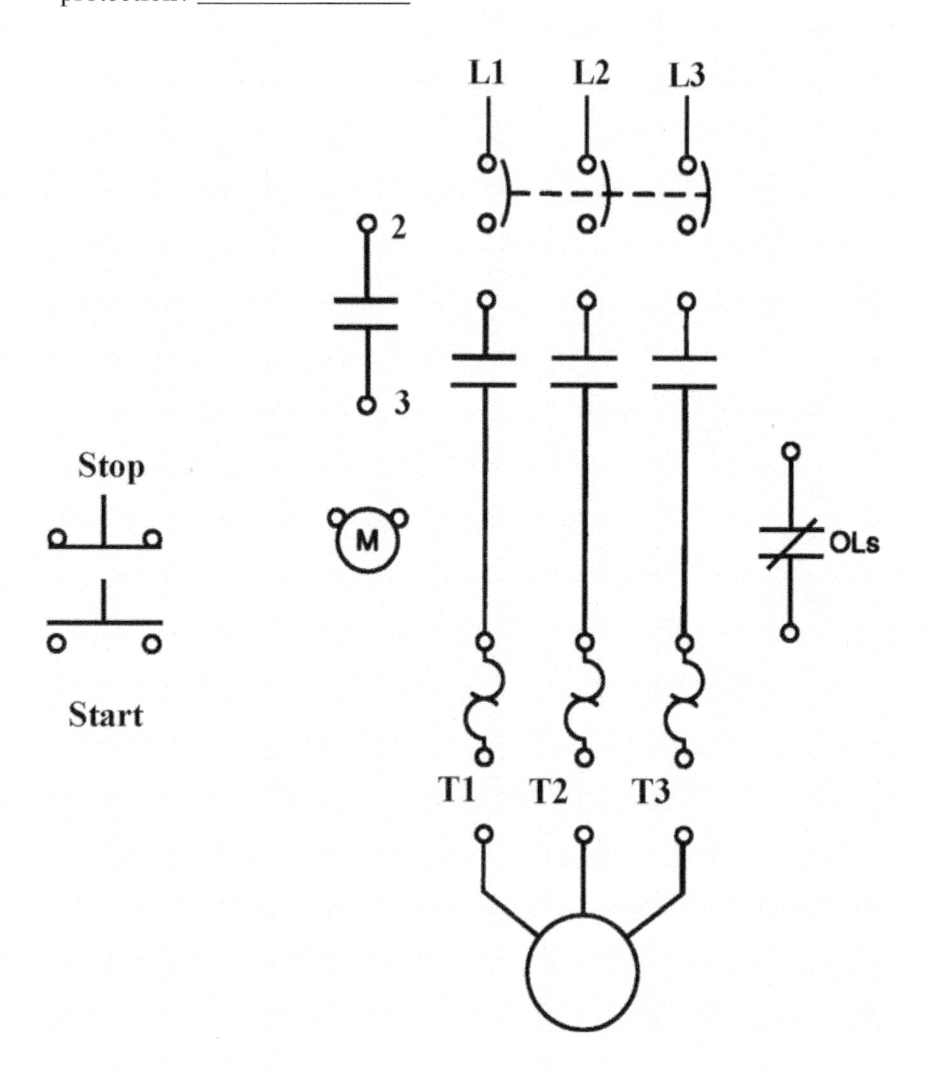

Figure 2-14 Three-wire control circuit for Practical Assignment 6.

3 Motor Transformers and Distribution Systems

PART 1 Quiz: Power Distribution Systems

Place the answers in the space provided.

1. The _______________ station of power generation and distribution enables power to be produced at one location for immediate use at other distant locations.
2. Without the use of _______________ to efficiently raise and lower voltage levels, the widespread distribution of electrical power would be impractical.
3. High voltages are used in transmission lines to reduce the transmission _______________.
   a. Resistance
   b. Current
   c. Power
   d. Power factor

   Answer _______
4. Large diameter conductors are used in electrical utility transmission systems relative to the power transmitted. (True/False) _______________
5. The higher the level of the distribution voltage, the more difficult and expensive it becomes to provide for adequate insulation of the conductors. (True/False) _______________
6. Three-phase power is usually supplied to commercial and industrial customers. (True/False) _______________

7. For the distribution system shown in Figure 3-1, determine the value of the transmission current. Answer _______

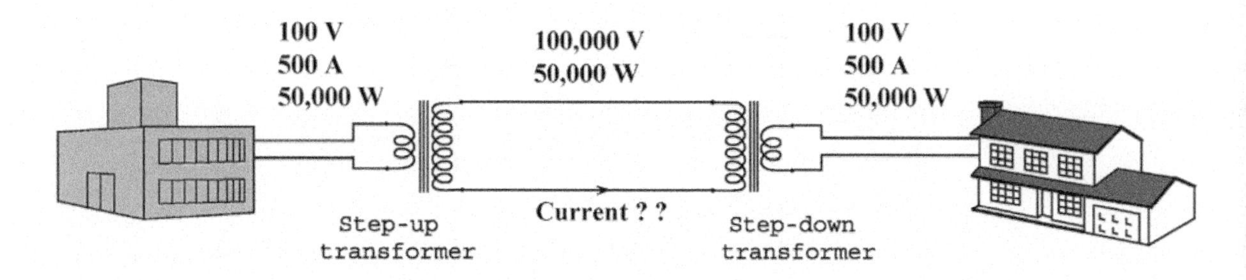

Figure 3-1 Circuit diagram for Question 7

8. Compared to the low-voltage side, the high-voltage side of a power transformer draws _______________ current.
   a. More
   b. Less
   c. The same
   d. A fixed amount of

   Answer _______

9. Substations contain _______________ transformers that reduce the transmission voltage levels.
10. The three major sections of the unit substation are
    a. ____
    b. ____
    c. ____
11. The unit substation is left open on the back for convenient access to all parts. (True/False) _______________
12. Before attempting to open the primary switch on a unit substation, you should disconnect the loads from the transformer. (True/False) _______________
13. Electrical distribution systems within buildings are required to safely deliver electrical energy without any component overheating or unacceptable voltage drops. (True/False) _______________

14. Identify the three major sections of the electrical distribution system shown in Figure 3-2.

a. ______________________________

b. ______________________________

c. ______________________________

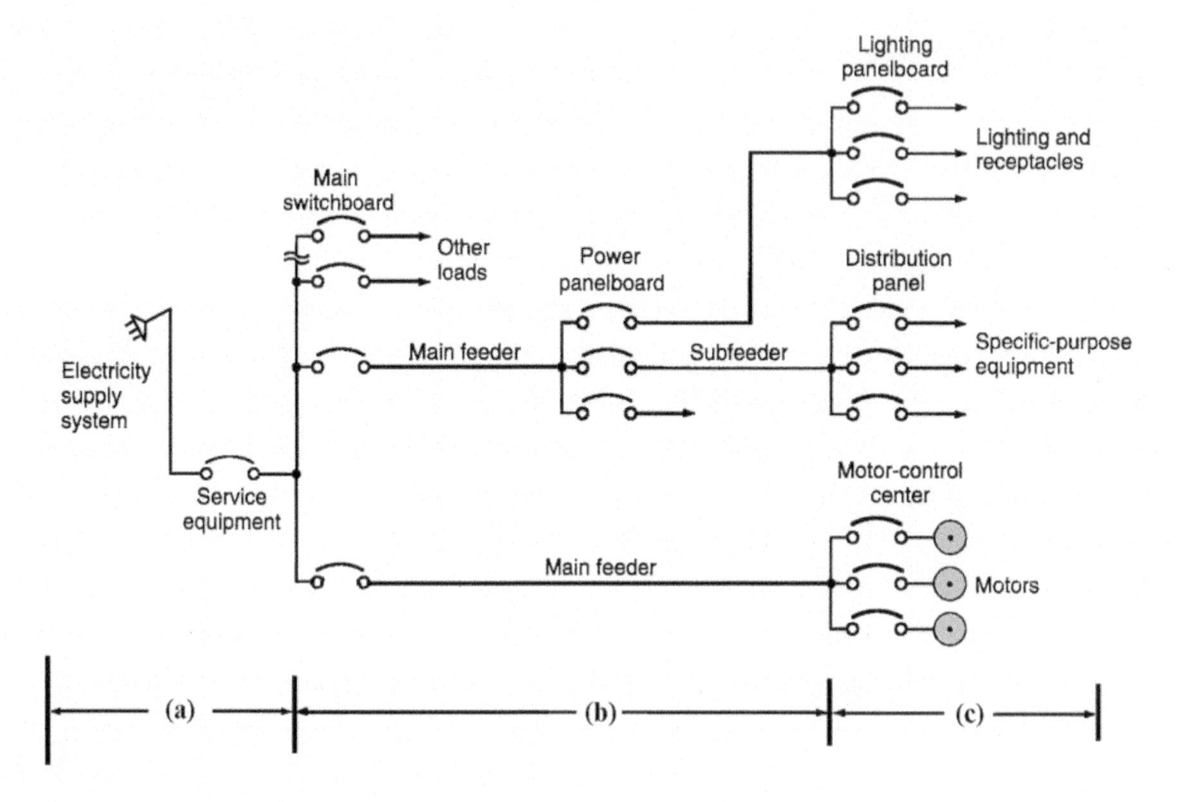

Figure 3-2 Electrical distribution system for Question 14.

15. Correct selection of conductors for feeders and branch circuits must take into account ______________, ______________, and ______________ requirements.

16. In installing motors and motor controllers, which of the following needs to be adhered to?

a. NEC

b. State and local codes

c. Manufacturers' instructions

d. All of these

Answer _______

17. The unit of electric energy generated by the power station always matches with that distributed to customers. (True/False) ____________

18. A power line of resistance R causes a power loss of

    a. $V \times I$

    b. $V^2 \times R$

    c. $I^2 \times R$

    d. $I \times R$

    Answer ______

19. Power line losses due to poor power factor can be reduced by the application of ____________ into the system.

20. Identify the three types of raceways shown in Figure 3-3.

    a. ____

    b. ____

    c. ____

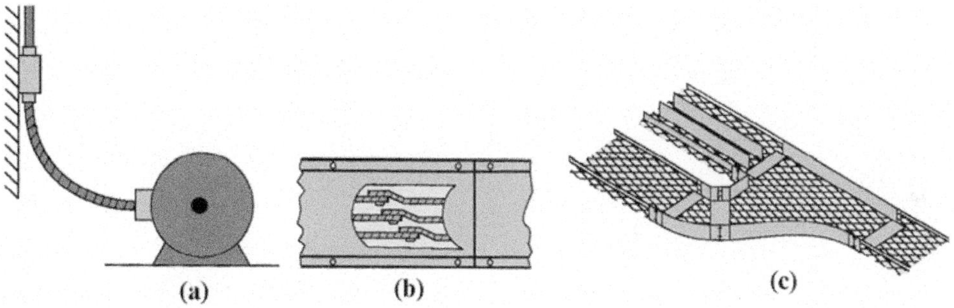

Figure 3-3 Raceways for Question 20.

21. Conduit capacity is generally based on a ____________ fill ratio.

    a. 90 percent

    b. 80 percent

    c. 60 percent

    d. 40 percent

    Answer ______

22. A panelboard is usually supplied from a switchboard and further divides the power distribution system into smaller parts. (True/False) ____________

23. What is the value of the line-to-neutral voltage for the three-phase four-wire panelboard feeder circuit shown in Figure 3-4?

Answer _______

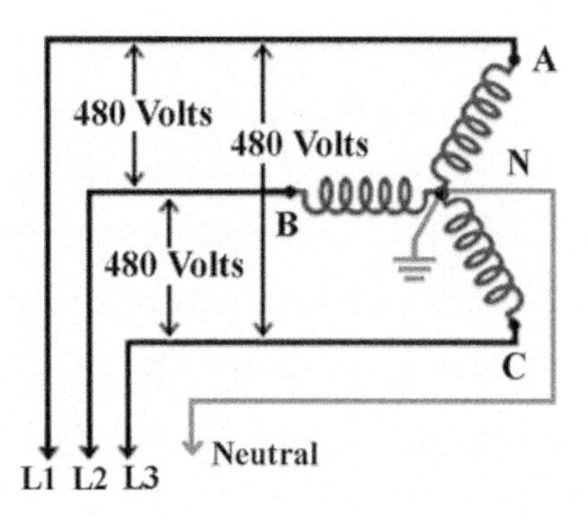

Figure 3-4 Feeder circuit for Question 23.

24. The proper grounding and bonding of an electrical distribution system ensures that any person who comes in contact with any _______________ parts of the installation will not receive an electric shock.
25. The impedance of the ground path is kept to a maximum. (True/False) _______________
26. The equipment grounding bus is noninsulated and connects directly to the panelboard metal enclosure. (True/False) _______________
27. Correctly identify the three-phase buses (A-B-C) for the panel bus arrangement shown in Figure 3-5.

(i) _______________ (ii) _______________ (iii) _______________

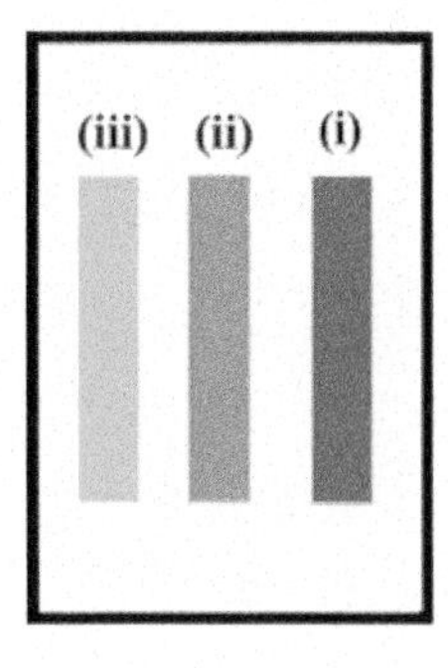

Figure 3-5 Panel bus arrangement for Question 27.

28. Main lug type panelboards contain a main breaker that is an integral part of the panelboard. (True/False) ______________
29. A motor control center is normally designed to accommodate plug-in–type motor control units. (True/False) ______________
30. A 3-phase full voltage starter contained within a motor control center consists of a ______________ and overload relay.
31. System grounding conductors are solidly connected to earth so that any ______________ current will safely flow to ground.
32. The equipment grounding is designed to protect personnel from electric shock. (True/False) ______________
33. ______________ grounding involves grounding of the noncurrent-carrying conductive part of electrical equipment.
    a. System
    b. Fused
    c. Equipment
    d. Circuit breaker

PART 2 Quiz: Transformer Principles

1. Identify the components of the transformer circuit shown in Figure 3-6.
   a. ______________ b. ______________
   c. ______________ d. ______________
   e. ______________ f. ______________

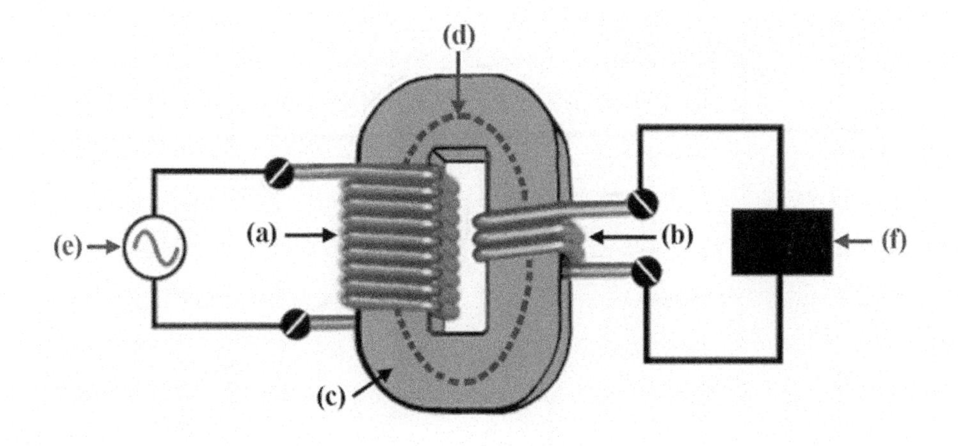

Figure 3-6 Transformer circuit for Question 1.

2. The principle of operation of a transformer is based on _______________ induction.
   a. Mutual
   b. Self
   c. Primary
   d. Secondary

   Answer _______
3. Movement of coils within a transformer produces the transformation of voltage. (True/False) _______________
4. For an ideal transformer, the power input is equal to the power output. (True/False) _______________
5. Transformer power is rated in volt-amperes instead of watts. (True/False) ______________
6. The voltage ratio of a transformer is equal to its turns ratio. (True/False) _______________
7. A step-up transformer is one in which the secondary winding current is greater than the primary winding current. (True/False) _______________
8. For the transformer circuit shown in Figure 3-7 determine the:
   a. Turns ratio ______________
   b. Value of the secondary voltage ______________

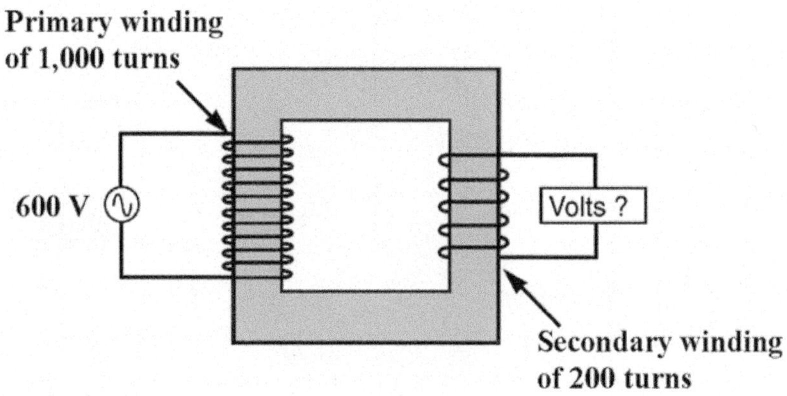

Figure 3-7 Transformer circuit for Question 8.

9. The efficiency of a transformer is typically in the ______________ range.

   a. 90 percent

   b. 80 percent

   c. 60 percent

   d. 40 percent

   Answer _______

10. When the secondary winding of a transformer is disconnected from the load,

   a. Zero current flows in the primary

   b. Very little current flows in the primary

   c. The primary current remains the same as under load

   d. The primary current becomes very high

   Answer _______

11. If the secondary circuit of a transformer becomes overloaded, the primary circuit will not be affected. (True/False) ______________

12. A transformer is being designed to increase the voltage from 12 V to 120 V. If the primary requires 400 turns of wire, how many turns are required on the secondary? ______________

13. For the transformer circuit shown in Figure 3-8 determine the

   a. Value of the secondary current flow ______________

   b. Value of the primary current flow ______________

   c. Value of the primary voltage ______________

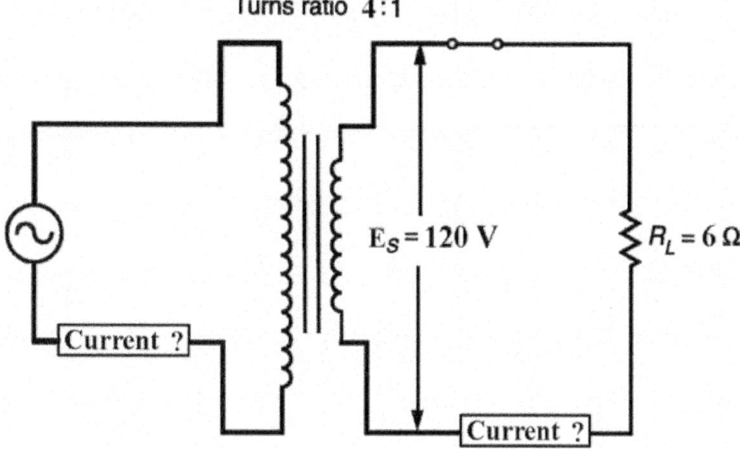

Figure 3-8 Transformer circuit for Question 13.

14. The ______________ of a transformer is a measure of the proportion of the applied energy that is transferred to the load.
15. Power losses associated with a transformer are generally relatively small. (True/False) ______________
16. The output secondary voltage of a transformer remains constant from no-load to the full-load condition. (True/False) ______________
17. Magnetizing inrush current in transformer
    a. Occurs when the transformer is first energized
    b. Is typically many times the normal full-load current
    c. Can result in nuisance tripping of the input overcurrent protection device
    d. All of these

    Answer _______

PART 3 Quiz: Transformer Connections and Systems

1. The high-voltage windings on a single-phase transformer are identified as X1 and X2. (True/False) ______________
2. On a transformer, when the lead marked H1 is instantaneously positive, the lead marked X1 will be instantaneously ______________.
3. A transformer has subtractive polarity when terminal H1 is adjacent to terminal X1. (True/False) ______________
4. What is the most popular operating voltage of commercial and industrial motor control systems?
    a. 120 V
    b. 208 V
    c. 440 V
    d. 600 V

    Answer _______

5. Correctly connect the dual-voltage control transformer circuits shown in Figure 3-9 for the input voltage levels given.

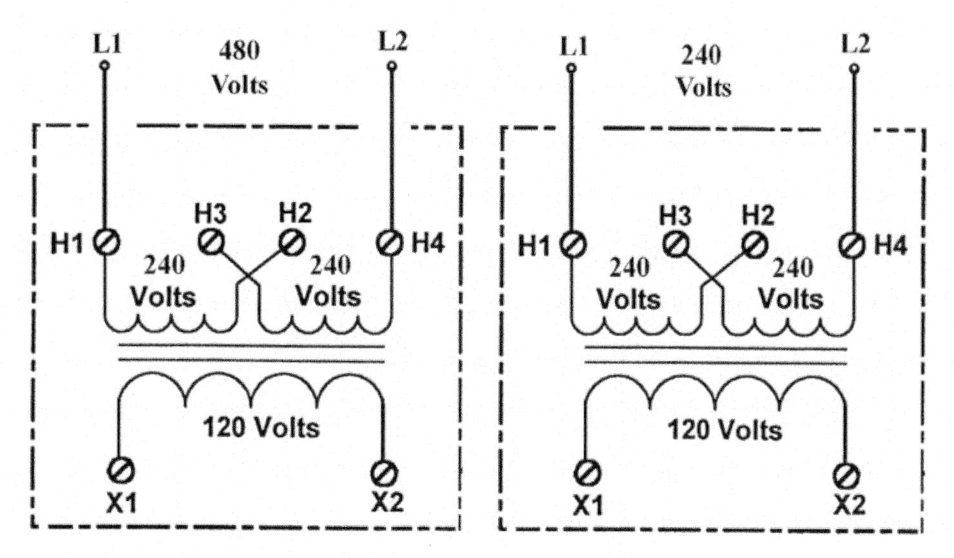

Figure 3-9 Dual-voltage transformer circuits for Question 5.

6. The primary of the three-phase transformer connection shown in Figure 3-10 is connected in a _______________ configuration and the secondary is connected in a _______________ configuration.

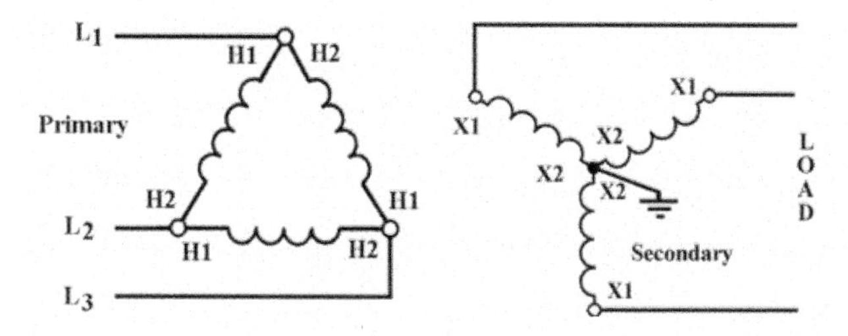

Figure 3-10 Three-phase transformer connection for Question 6.

7. In three-phase transformer systems, the constant _______________ is used because the transformer phase windings are 120 electrical degrees apart.

   a. 3

   b. 2.5

   c. 1.73

   d. 0.707

   Answer _______

8. In a Delta-connected transformer secondary, the phase and line voltages are equal. (True/False) ______________

9. For the Wye distribution system shown in Figure 3-11, if the voltage between any two line leads is 208 V, the voltage from any line lead to neutral would be ______________ V.

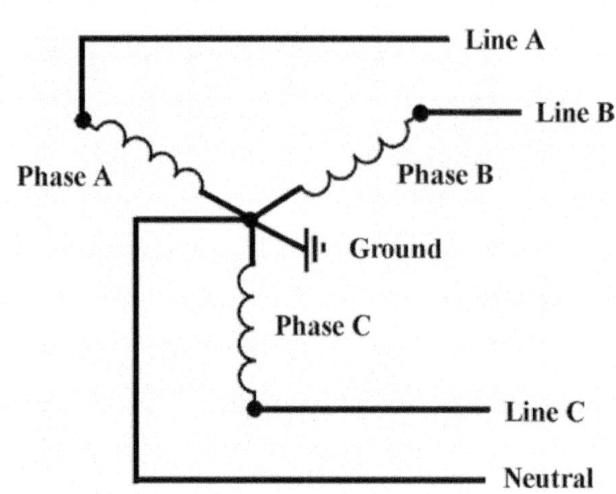

Figure 3-11 Wye distribution system for Question 9.

10. An autotransformer has electrical isolation between the primary and secondary circuits. (True/False) ______________

11. An autotransformer motor starter reduces the starting current to the motor by
    a. Inserting resistance in series with the motor
    b. Reducing the value of the applied voltage
    c. Reducing the load on the motor
    d. Reducing the frequency of the applied voltage

    Answer _______

12. Compared to an equivalent traditional transformer, the autotransformer is smaller in size. (True/False) ______________

13. The secondary low-voltage side of a potential instrument transformer is usually wound for
    a. 5 volts
    b. 10 volts
    c. 24 volts
    d. 120 volts

    Answer _______

14. For the circuit of Figure 3-12, transformer X is a _____________ transformer, while transformer Y is a _____________ transformer.

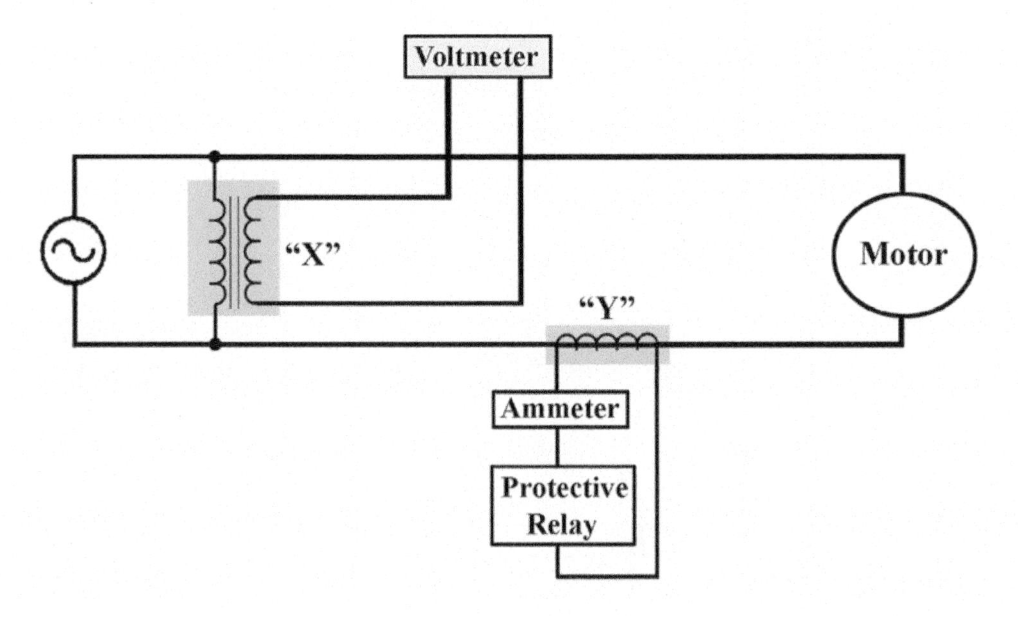

Figure 3-12 Circuit for Question 14.

15. Instrument transformers provide insulation between the instrument and the high voltage of the power circuit. (True/False) _____________

16. A current transformer should always have its secondary shorted when there is current flow in the primary winding but no load connected to the secondary. (True/False) _____________

17. An insulation check of transformer windings is made using a _____________.

    a. ammeter

    c. ohmmeter

    b. megger

    d. voltmeter

18. When making an insulation test of a transformer winding both the resistance insulation to ground as well as between coil windings should be tested. (True/False) _____________

hands on PRACTICAL ASSIGNMENTS

1. a. The purpose of this assignment is to connect a magnetic across-the-line motor starter to operate a three-phase motor using start/stop push buttons and a control transformer. Complete the wiring diagram for the magnetic starter circuit shown in Figure 3-13.

   b. Wire the circuit in a neat and professional manner.

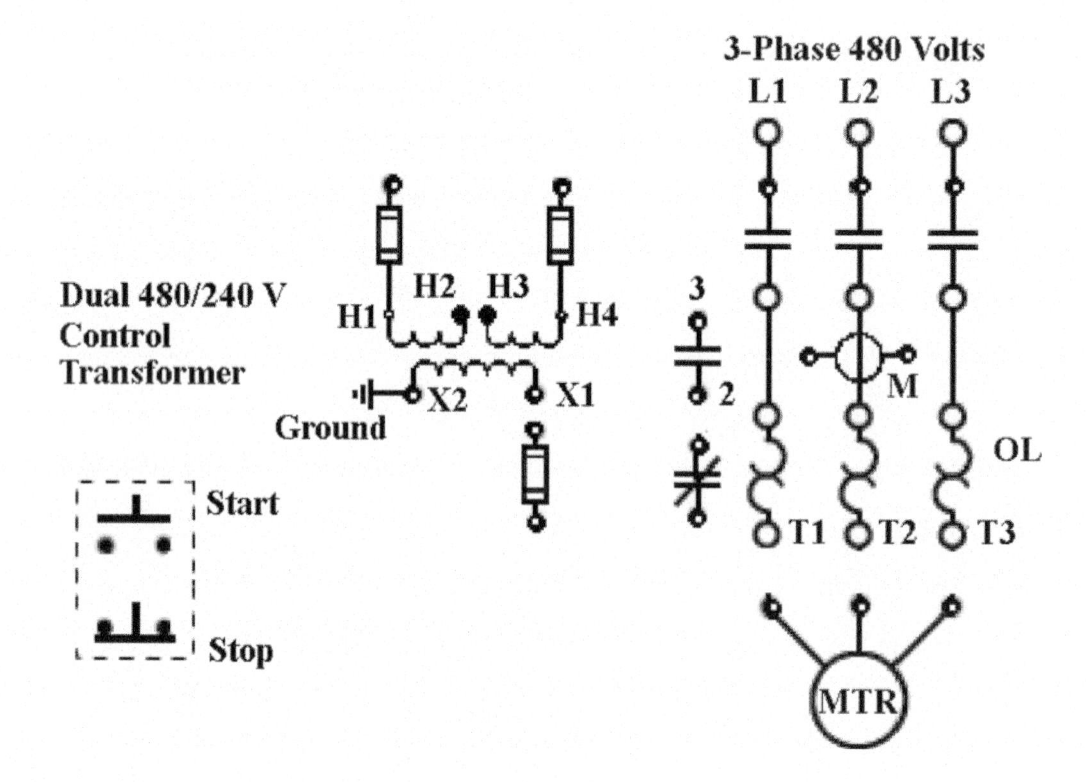

Figure 3-13 Control transformer circuit for Practical Assignment 1.

2. The purpose of this assignment is to develop a wiring diagram for a typical Delta-to-Wye transformer bank connection. Connect the primary transformer lines to the distribution system to form the Delta-connected primary of Figure 3-14. Connect the secondary transformer lines to the distribution system to form a Wye-connected secondary that provides 208-V three phase, 208-V single phase, and 120-V single phase. Connect each motor load to the correct power supply.

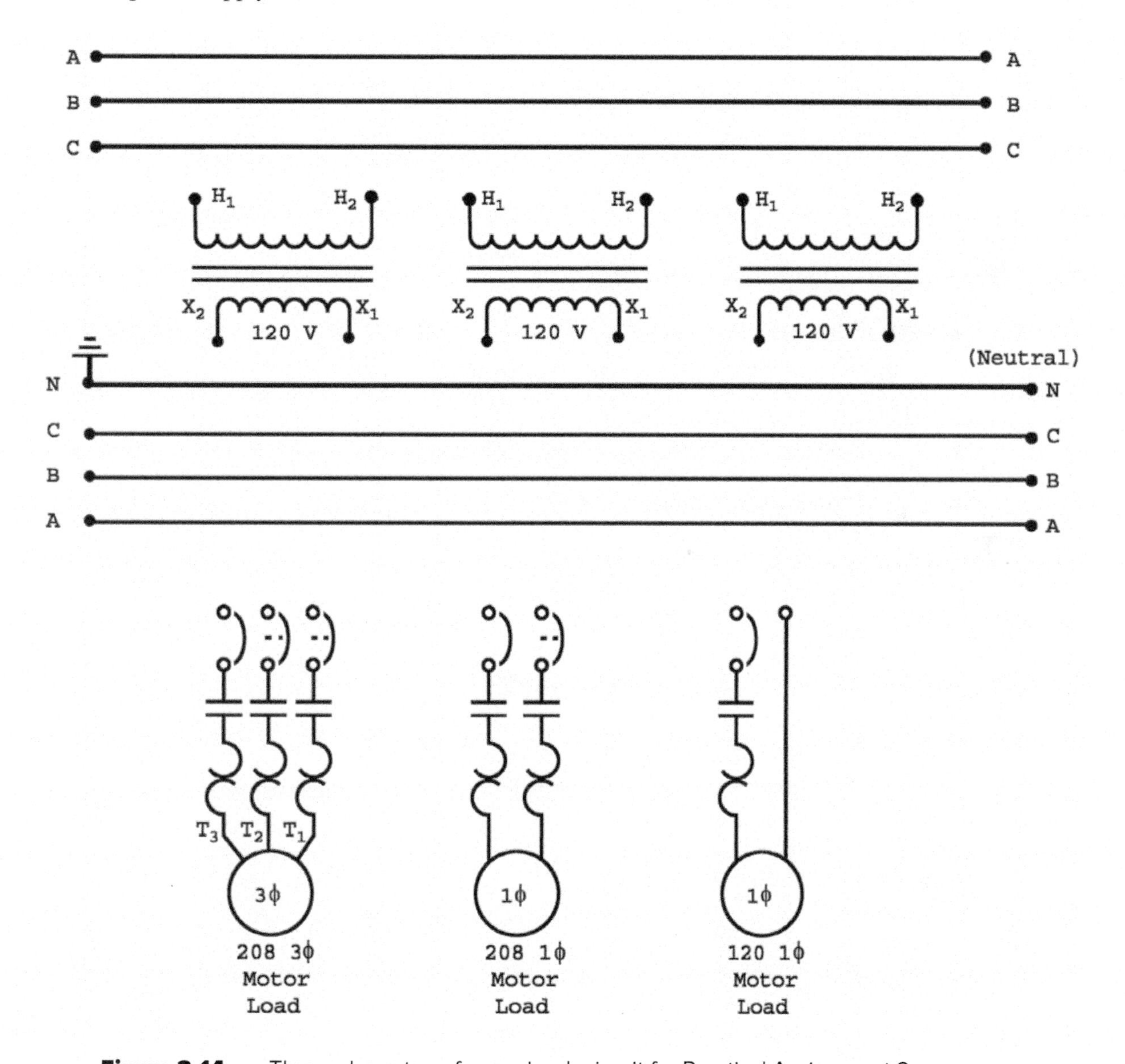

Figure 3-14 Three-phase transformer bank circuit for Practical Assignment 2.

4 Motor Control Devices

PART 1 Quiz: Manually Operated Switches

Place the answers in the space provided.

1. Pilot-duty devices are normally used to switch horsepower-rated loads. (True/False) _______________
2. Which of the following would be classified as a primary control device?
   a. Flow switch
   b. Thermostat
   c. Motor starter
   d. Pressure switch

   Answer _______
3. A push button would normally be considered to be a pilot control device. (True/False) _______________
4. A manually operated switch is one that is controlled by _______________.
5. The abbreviation SPDT stands for a _______________ _______________ _______________ _______________ switch.
6. The type of switch shown in Figure 4-1 is a _______________.

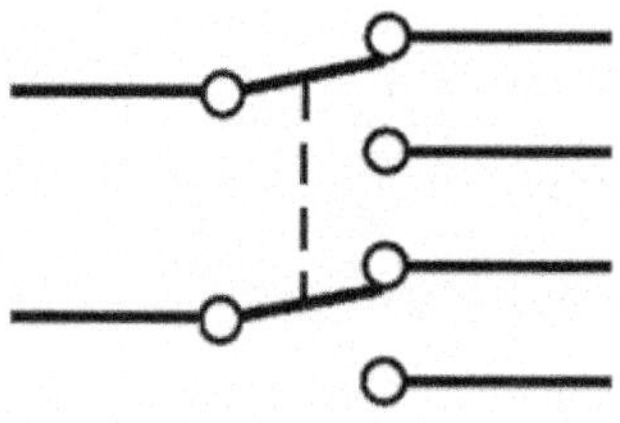

Figure 4-1 Switch for Question 6.

7. The DC current rating of a switch would have a lower magnitude than the AC rating. (True/False) _______________

8. Identify each type of push button shown in Figure 4-2 and draw the NEMA symbol used to represent it.

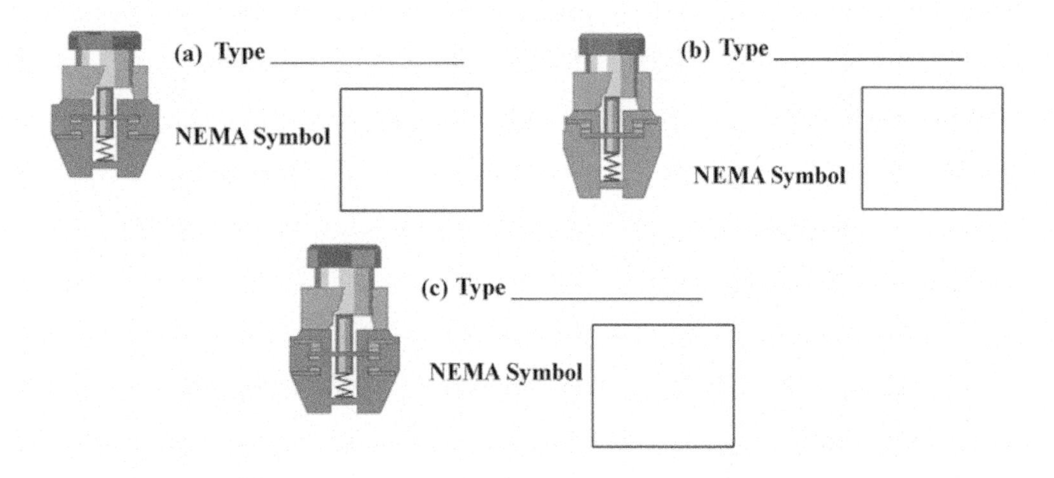

Figure 4-2 Push buttons for Question 8.

9. The NC push button _______________ the circuit when it is pressed and returns to the _______________ position when the button is released.
10. The break-make push button is used for ____________________ control circuits.
11. Electrical _______________ are designed to protect their contents from operating environmental conditions.
12. A push button may consist of one or more _______________ blocks, a(n) _______________ device, and a(n) _______________ plate.
13. _______________ head pushbutton operators are used for emergency stop push buttons.
14. A pushbutton assembly can only have a one contact block. (True/False) _______________
15. Once actuated, _______________ type pushbutton operators are required to be actuated a second time to return the contacts to their normal off state.
16. Emergency stop push buttons must be of the maintained contact type. (True/False) _______________

17. Draw the NEMA and IEC symbols used to represent a red pilot light.

NEMA IEC

18. Push-to-test pilot lights can be energized from two separate signals of different voltage values. (True/False) _______________

19. All selector switches operate with spring return to provide momentary contact operation. (True/False) _______________

20. Drum switches are used for starting and reversing three-phase squirrel-cage motors. (True/False) _______________

21. In a multi-function process control system the emergency stop function must be available and operational at all times, regardless of the operating mode. (True/False) _______________

22. Light towers are designed to provide visual indicators of machine ____________ to factory personnel.

PART 2 Quiz: Mechanically Operated Switches

1. A mechanically operated switch is one that is controlled by hand. (True/False) _______________

2. Limit switches are designed to be operated:
   a. By the machine operator
   b. When a predetermined pressure is reached
   c. When a predetermined temperature is reached
   d. By contact with an object

   Answer _______

3. Identify each of the switch symbols shown in Figure 4-3.

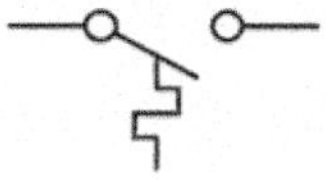

a) ____________________ **b)** ____________________ **c)** ____________________

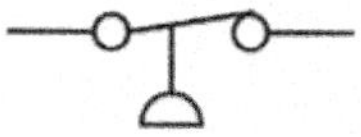

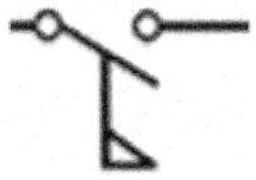

d) ____________________ **e)** ____________________

Figure 4-3 Switch symbols for Question 3.

4. The type of limit switch shown in Figure 4-4 would be classified as a

   a. Lever type

   b. Push roller type

   c. Wobble-stick type

   d. Fork lever type

   Answer _______

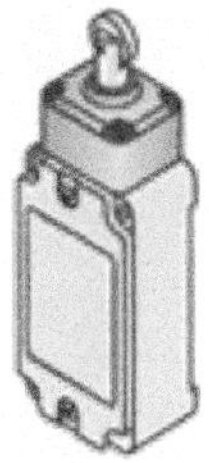

Figure 4-4 Limit switch for Question 4.

5. Micro limit switches require a large amount of pressure to be applied to the operating lever in order to actuate the contacts. (True/False) _______________

6. The type of temperature switch shown in Figure 4-5 is

   a. Bimetallic type

   b. Capillary type

   c. Float type

   d. Spring type

   Answer _______

Figure 4-5 Temperature switch for Question 6.

7. All temperature switches use only N.O. (normally open) contacts. (True/False) _______________

8. The three categories of pressure switches are ______________ pressure, ______________ pressure, and ______________ pressure.

9. Pressure switches open or close when a preset pressure is reached. (True/False) ______________

10. The type of switch shown in Figure 4-6 would be classified as a

    a. Pressure type

    b. Flow type

    c. Float type

    d. Temperature type

    Answer ___________

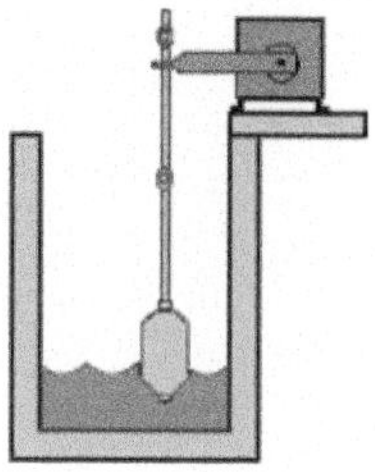

Figure 4-6 Switch for Question 10.

11. A flow switch is used to detect the ______________ of air or liquid.

12. Limit switches contain no moving mechanical parts. (True/False) ___________

13. One of the strengths of limit switches is that they can be used in most industrial environment due to their rugged design. (True/False) ______________

14. The part of a limit switch that physically comes in contact with the target is called the ___________________

PART 3 Quiz: Sensors

1. Sensors are devices that are used to ______________ and ______________ the presence of something.

2. A proximity sensor can detect the presence of an object without physical contact. (True/False) ______________

3. An inductive proximity sensor can be actuated by both conductive and nonconductive materials. (True/False) ______________

4. Identify each component block of the inductive proximity sensor shown in Figure 4-7.

a. ______________________ b. ______________________

c. ______________________ d. ______________________

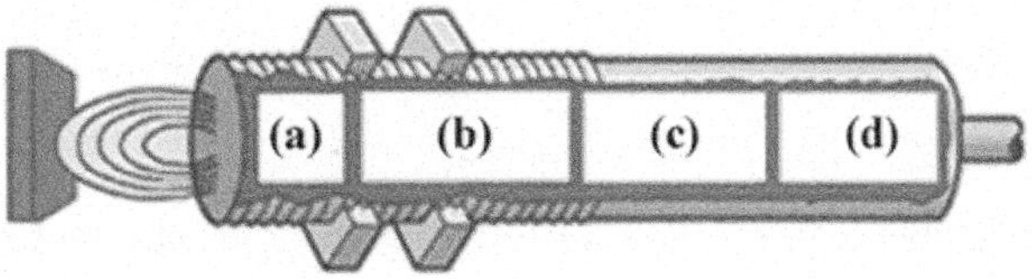

(a) (b) (c) (d)

Figure 4-7 Inductive proximity sensor for Question 4.

5. Draw acceptable NEMA and IEC symbols used to represent an N.O. proximity switch.

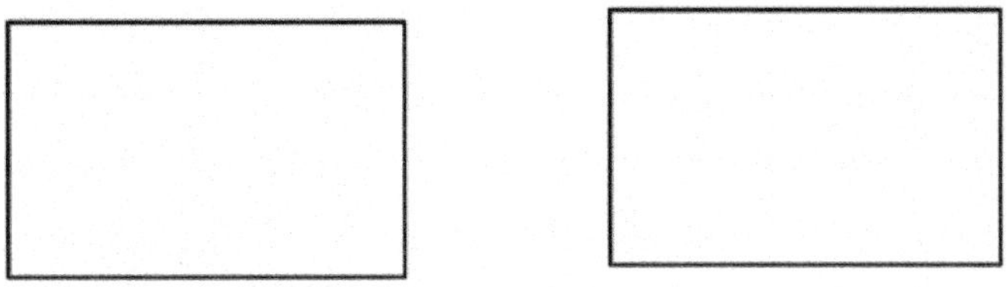

NEMA IEC

6. The proximity sensor shown in Figure 4-8 would be classified as a

   a. Three-wire type
   b. Two-wire type
   c. DC type
   d. Both a and c

   Answer _______

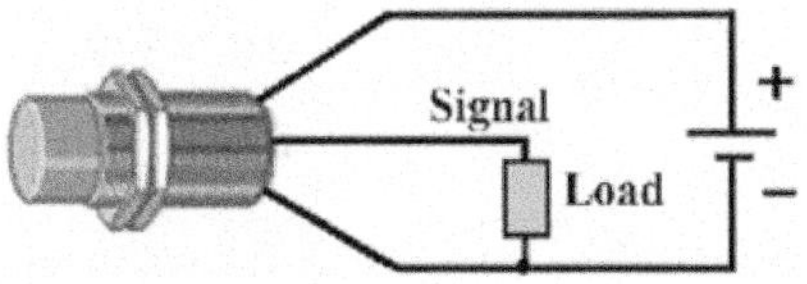

Signal
Load
+
−

Figure 4-8 Proximity sensor for Question 6.

7. A proximity sensor should not normally be used to directly operate a motor. (True/False) ______________

8. The electronic switching circuit of a sensor

   a. Has infinite resistance in the off state
   b. Will allow a small amount of current to flow through it in the off state
   c. Will have a small voltage drop across it in the on state
   d. Both b and c

9. When a metal object enters the high-frequency field of an inductive proximity sensor, it causes
   a. A change in amplitude of the oscillating circuit
   b. The oscillator circuit to start oscillating
   c. Vibration of the oscillator
   d. Both b and c

   Answer _______
10. In order to operate properly, a two-wire proximity sensor should be powered continuously. (True/False) _______________
11. The area between the operating and release points of a proximity sensor is known as the _______________ zone.
12. The position identified on the field of the proximity sensor shown in Figure 4-9 is known as the
   a. Saturation point
   b. Hysteresis point
   c. Operating point
   d. Release point

   Answer _______

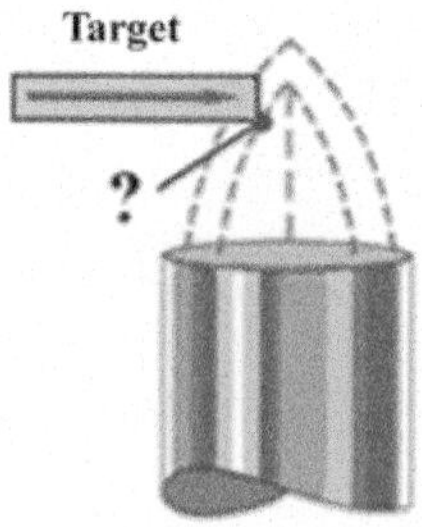

Target

?

Figure 4-9 Proximity sensor for Question 12.

13. A capacitive proximity sensor produces a(n)
   a. Electromagnetic sensing field
   b. Electrostatic sensing field
   c. Constant current sensing field
   d. Constant voltage sensing field

   Answer _______
14. When the target enters the field of a capacitive proximity sensor, the oscillator circuit begins oscillating. (True/False) _______________
15. Capacitive proximity sensors will sense only nonmetallic materials. (True/False) _______________

16. Identify the two basic components of the photoelectric sensor shown in Figure 4-10.

a. ____________________________

b. ____________________________

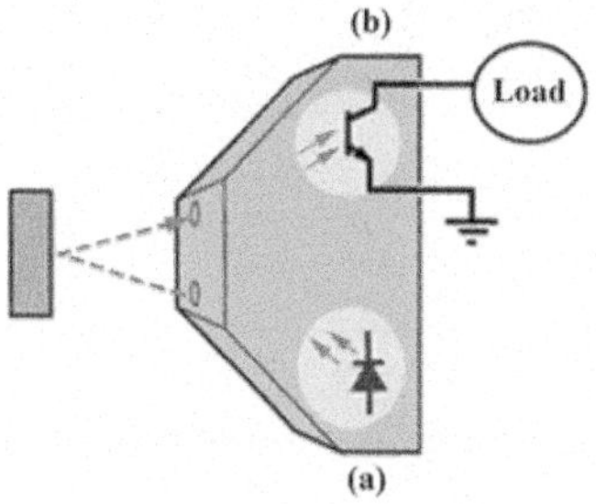

Figure 4-10 Photoelectric sensor for Question 16.

17. Identify the type of scan technique used for each of the photoelectric sensors shown in Figure 4-11.

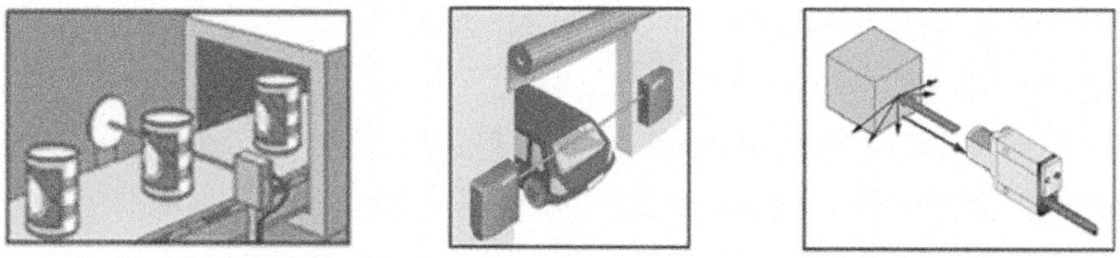

(a) ____________________ (b) ____________________ (c) ____________________

Figure 4-11 Photoelectric sensors for Question 17.

18. Which of the following scan techniques is best suited for long-range scanning?

   a. Through-beam

   b. Retroreflective

   c. Polarized retroreflective

   d. Diffuse

   Answer _______

19. A ______________ retroreflective scan sensor is designed to be able to detect shiny objects.

20. Fiber optics can be used with through-beam, retroreflective, or diffuse scan sensors. (True/False) ______________

21. Fiber optic sensor systems are completely immune to all forms of electrical interference. (True/False) ______________

22. Hall-effect sensors are used detect both the proximity and the ______________ of a magnetic field.
23. Identify the measurement application for the Hall-effect sensing circuits shown in Figure 4-12.
    a. Application:
    ____
    ____
    ____
    b. Application:
    ____
    ____
    ____

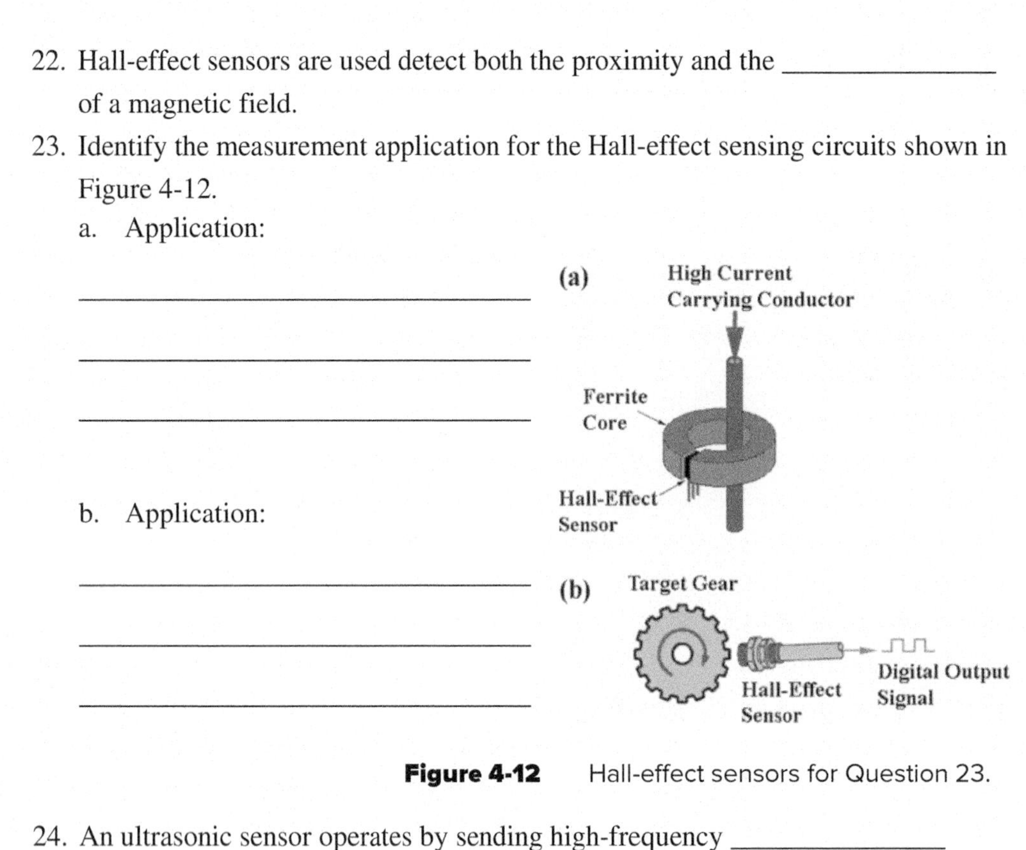

Figure 4-12 Hall-effect sensors for Question 23.

24. An ultrasonic sensor operates by sending high-frequency ______________ waves toward the target and measuring the ______________ it takes for the pulses to bounce back.
25. For the ultrasonic sensor shown in Figure 4-13, the sensor will generate a signal of approximately ______________ mA when the tank is full and ______________ mA when the tank is empty.

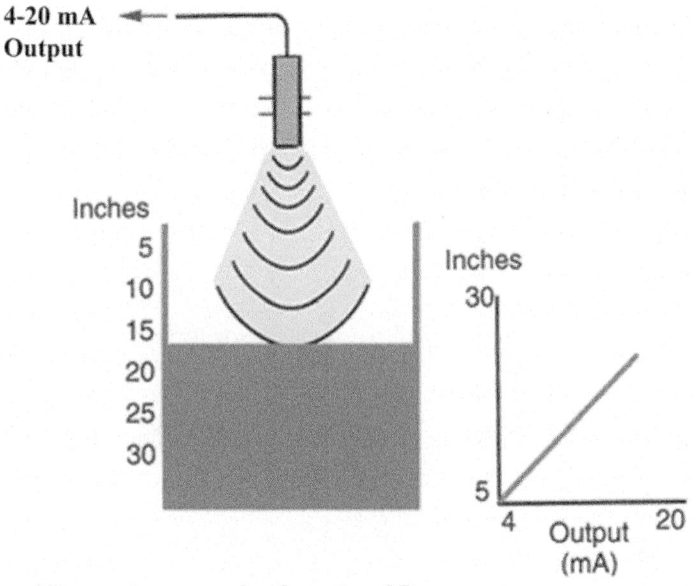

Figure 4-13 Ultrasonic sensor for Question 25.

26. Identify each type of temperature sensor shown in Figure 4-14.

(a) ____________ (b) ____________ (c) ____________ (d) ____________

Figure 4-14 Temperature sensors for Question 26.

27. Which of the following temperature sensors operates on the principle that electrical resistance of metals is directly proportional to temperature?
    a. IC
    b. RTD
    c. Thermocouple
    d. Thermistor

    Answer _______

28. The output voltage of a thermocouple is proportional to the difference in voltage between the ______________ junction and the ______________ junction.

29. Different thermocouple types basically have the same voltage output curves. (True/False) ______________

30. Which of the following temperature sensors provides the most accurate measurement of temperature?
    a. IC
    b. RTD
    c. Thermocouple
    d. Thermistor

    Answer _______

31. The material most often used for resistance temperature detectors is
    a. Platinum
    b. Gold
    c. Copper
    d. Aluminum

    Answer _______

32. As the temperature of a thermistor with a negative temperature coefficient increases, its resistance decreases. (True/False) _______________

33. The operating temperature range of the IC temperature sensor is much lower than that of a thermocouple. (True/False) _______________

34. _______________ are tubular fittings used to protect temperature sensors installed in industrial processes.

    Answer _______

35. A tachometer generates a voltage that is proportional to _______________.

36. Identify the parts of the magnetic pickup sensor shown in Figure 4-15.

    a. ____________________________

    b. ____________________________

    c. ____________________________

    d. ____________________________

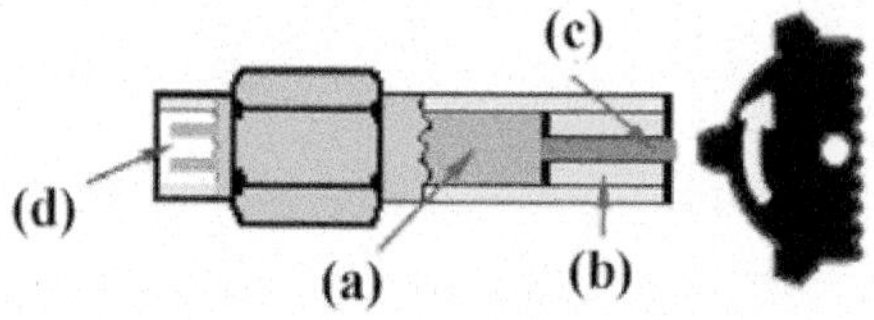

Figure 4-15 Magnetic pickup for Question 36.

37. A magnetic pickup sensor generates pulses that are counted to determine shaft speed. (True/False) _______________

38. An encoder sensor is used to convert linear or rotary motion into a binary _______________ signal.

39. Encoder sensors are used in applications such as robotic control where _______________ have to be accurately determined.

40. Identify each type of flowmeter shown in Figure 4-16.

    a. ____________________________

    b. ____________________________

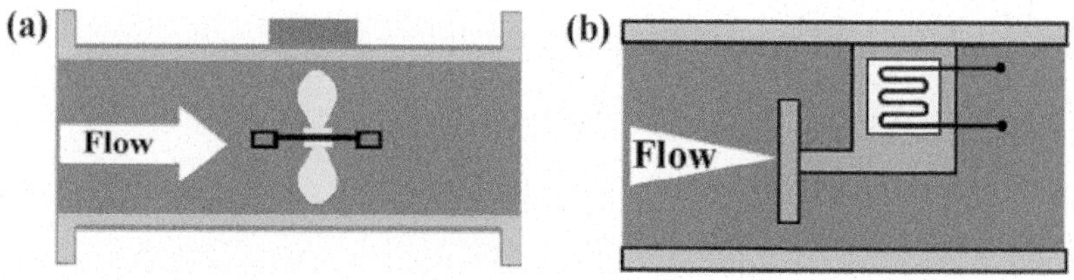

Figure 4-16 Flowmeters for Question 40.

41. The usual approach taken to measure fluid flow is to convert the ______________ energy that the fluid has into some other measurable form.

42. Which of the flowmeters operates by measuring the changes of induced voltage of a conductive fluid?
   a. Magnetic
   b. Target
   c. Turbine
   d. Differential pressure

   Answer _______

43. Inductive proximity sensors can detect metallic targets through nonmetallic barriers. (True/False) ______________

44. Capacitive proximity sensor applications are able to detect both metallic and nonmetallic objects. (True/False) ______________

45. _______use light to detect the presence or absence of an object.
   a. Limit switches
   b. Proximity sensors
   c. Generators
   d. Photoelectric sensors

46. Which type of sensor is best for detecting liquid targets?
   a. Inductive proximity sensors
   b. Capacitive proximity sensors
   c. Limit switches
   d. Photoelectric sensors

47. Which of the following pressure switches contains a sealed chamber that has multiple ridges like the pleats of an accordion?
   a. Bourdon tube
   b. Selector
   c. Bellows
   d. Diaphragm

48. The type of pressure switch shown is
   a. Bourdon tube
   b. Selector
   c. Bellows
   d. Diaphragm

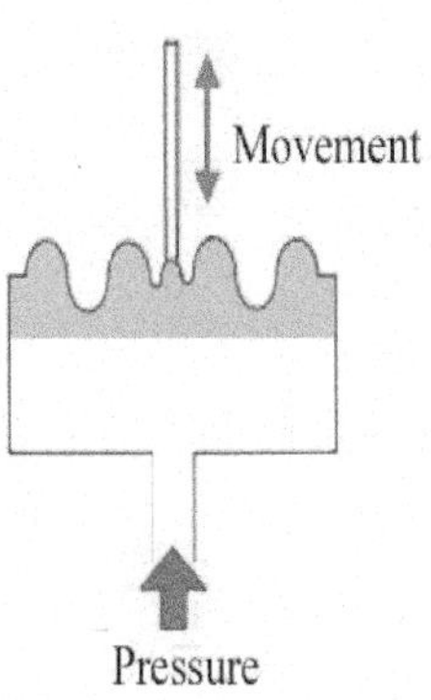

PART 4 Quiz: Actuators

1. An actuator is any device that converts an electrical signal into _______________ _______________.
2. The contacts of an electromechanical relay are switched by energizing and de-energizing a(n) _______________.
3. The normally closed contact of a relay is closed any time the coil is energized. (True/False) _______________.
4. Compared to a single-break contact, a double-break relay contact
   a. Uses two pair of contacts
   b. Can handle higher voltages
   c. Provides longer contact life
   d. All of these

   Answer _______
5. A dry contact usually means that it is completely isolated and has no voltage applied to it from its initiating equipment. (True/False) _______________
6. A solenoid is made up of coil with a movable _______________.
7. Which type of solenoid is shown in Figure 4-17?
   a. AC
   b. DC
   c. Rotary
   d. Manual

   Answer _______

Figure 4-17 Solenoid for Question 7.

8. An AC solenoid rated for 120-V alternating current, if connected to a 120-V direct current source, would draw a much lower than normal current. (True/False) _______________.
9. If a DC solenoid sticks in the open position, a burnout of the coil is likely. (True/False) _______________.
10. In AC solenoids, coil current decreases as the plunger moves toward the seated position. (True/False) _______________.

11. A solenoid valve is a combination of a solenoid coil operator and
    a. Contacts
    b. Sensor
    c. Transmitter
    d. Valve

    Answer _______

12. When a solenoid valve is electrically energized it can be designed to
    a. Open the flow of media
    b. Shut off the flow of media
    c. Direct the flow of media
    d. All of these

    Answer _______

13. Stepper motors rotate continuously when voltage is applied to their terminals. (True/False) _______________.

14. For a DC stepper motor, the amount of rotation is directly proportional to the _______________ of pulses and the speed to the _______________ of the pulses.

15. Generally, stepper motors are rated for more than 1 hp and are used to operate very heavy loads. (True/False) _______________.

16. Stepper motor systems can provide precise position control of movement without the use of a feedback signal. (True/False) _______________.

17. An open-loop motor control system is one that is operated
    a. Without a feedback signal
    b. With a feedback signal
    c. Without a power signal
    d. With a power signal

    Answer _______

18. All servo motors operate using a closed-loop control system. (True/False) _______________.

19. Identify all blocks for the servo motor control system shown in Figure 4-18.

a. ______________________

b. ______________________

c. ______________________

d. ______________________

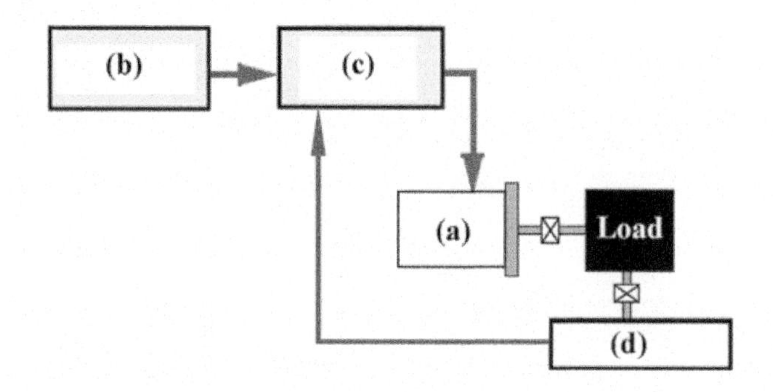

Figure 4-18 Servo system for Question 19.

20. Brushless DC motors are mechanically commutated.

(True/False) ______________.

hands on PRACTICAL ASSIGNMENTS

1. a. The purpose of this assignment is to connect pilot lights to a magnetic across-the-line motor starter circuit. Pilot lights are to be connected so that the green light is on when the starter coil is deenergized and the red light is on when the starter coil is energized. Complete the wiring diagram for the magnetic across-the-line motor starter circuit of Figure 4-19.

b. Wire the circuit in a neat and professional manner.

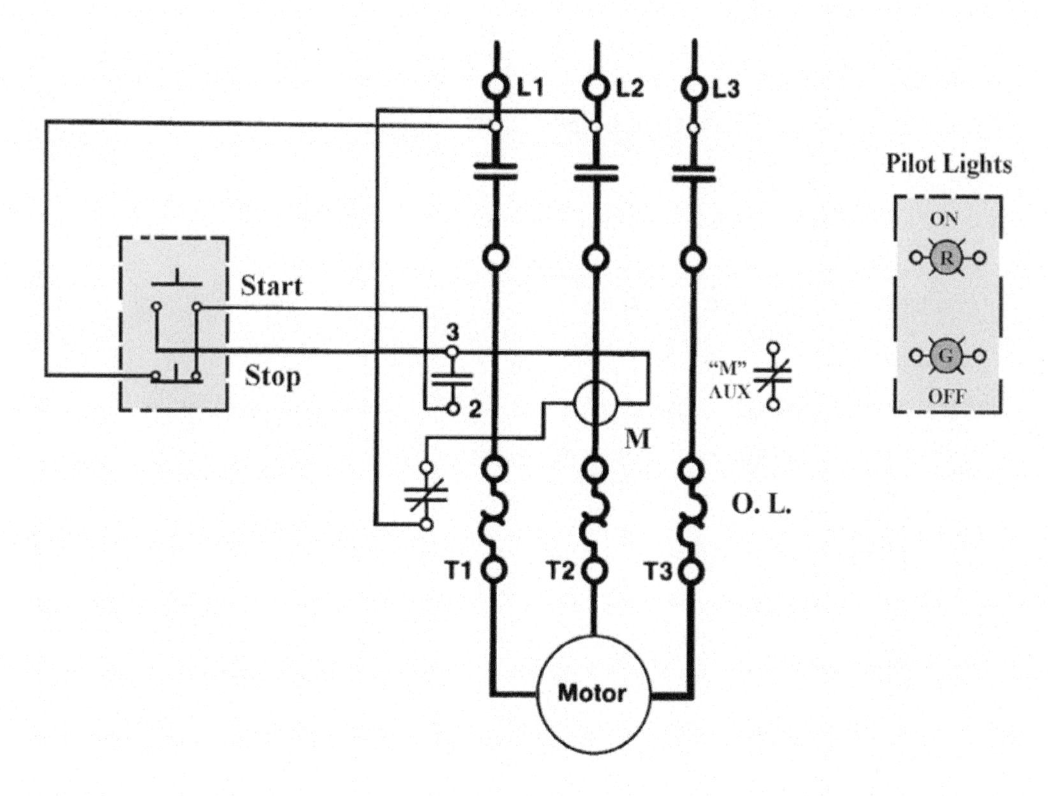

Figure 4-19 Magnetic across-the-line motor starter circuit for Practical Assignment 1.

5 Electric Motors

PART 1 Quiz: Motor Principle

Place the answers in the space provided.

1. Electric motors are used to convert _______________ energy into _______________ energy.
2. An electric motor uses _______________ and electric currents to operate.
3. The two basic categories of motors are _______________ and _______________.
4. The magnetic field of a magnet is represented by
   a. Positive charges
   b. Negative charges
   c. Lines of current
   d. Lines of flux

   Answer _______
5. Draw the shape and indicate the direction of the magnetic field for the conductor shown in Figure 5-1.

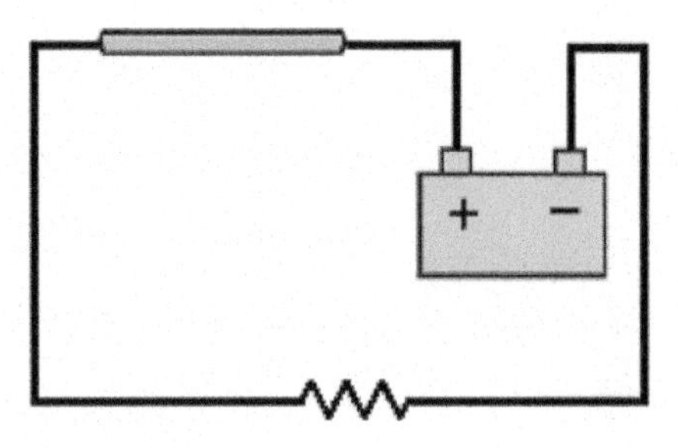

Figure 5-1 Circuit for Question 5.

6. The magnetic field produced by a current-carrying coil resembles that of a permanent magnet. (True/False) _______________
7. The polarity of the poles of a current-carrying coil reverses whenever the current flow through the coil reverses. (True/False) _______________

8. An electric generator is a machine that uses magnetism to convert ______________ energy into ______________ energy.
9. AC generators are also called ______________ because they produce an alternating current.
   a. Alternators
   b. Commutators
   c. Motors
   d. Rectifiers
10. A generator produces electricity by
    a. Aligning magnetic lines of force
    b. Increasing voltage
    c. Rotating a loop of wire in a magnetic field
    d. Decreasing resistance
11. The commutator of a DC generator converts the generated AC voltage into DC voltage. (True/False) ______________
12. An electric motor rotates as the result of the interaction of two magnetic fields. (True/False) ______________
13. Complete the drawing of Figure 5-2 showing
    - The direction of the electron flow through the conductor
    - The direction of the magnetic field of the permanent magnets
    - The direction in which the conductor would move

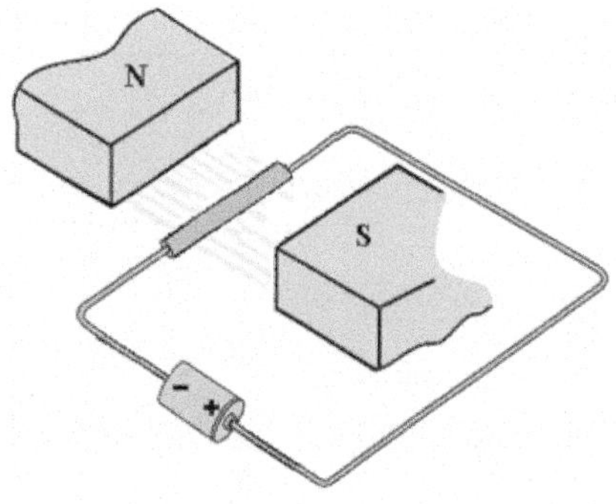

Figure 5-2 Circuit for Question 13.

PART 2 Quiz: Direct Current Motors

1. Direct current motors are used where a wide range of high torque and speed control is required. (True/False) _______________
2. Use of DC motors is limited by the fact that
   a. Electric utility systems deliver alternating current
   b. DC motor types are more expensive to construct than AC types
   c. DC motor types require more maintenance than AC types
   d. All of these

   Answer ________
3. Identify the major components of the DC motor shown in Figure 5-3.
   a. ______________________________
   b. ______________________________
   c. ______________________________
   d. ______________________________
   e. ______________________________
   f. ______________________________
   g. ______________________________
   h. ______________________________

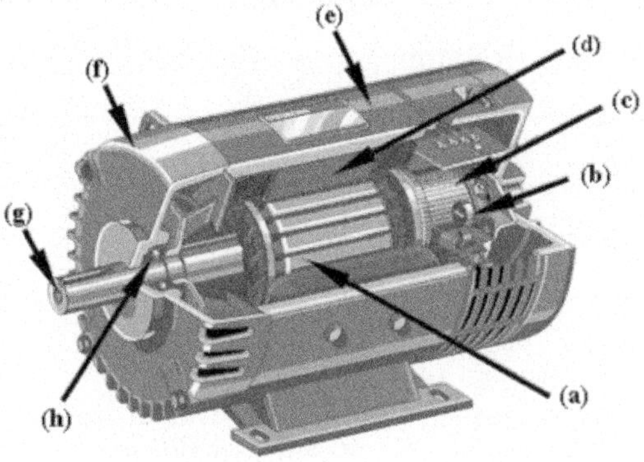

Figure 5-3 Motor for Question 3.

4. The rotational speed of a motor's shaft is usually measured in _______________.
5. _______________ refers to the turning force supplied by the motor's shaft.
6. One horsepower is the equivalent of _______________ watts of electrical power.
7. The action of switching current between coils within the armature of a DC motor is called _______________.
8. DC permanent-magnet motors are often used in servo motor applications. (True/False) _______________

9. Answer each of the following with reference to the DC motor shown in Figure 5-4.
    a. This motor would be classified as a ______________ type.
    b. The direction of rotation is changed by ______________

    ____
    c. The speed is varied by ____

    ____

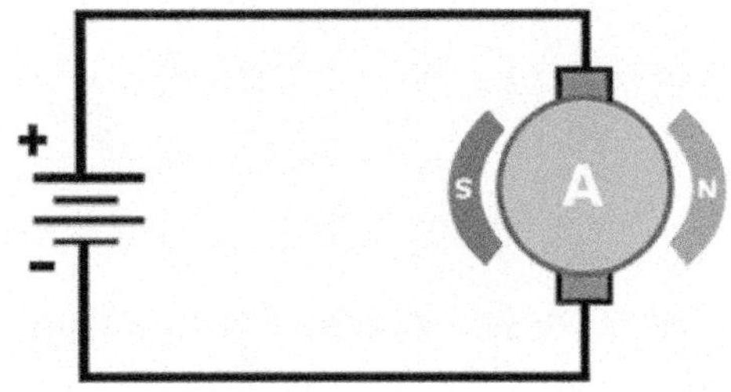

Figure 5-4 Motor circuit for Question 9.

10. A series-type DC motor uses field coils with
    a. Low resistance and made up of a few turns of large-diameter wire
    b. Low resistance and made up of a many turns of small-diameter wire
    c. High resistance and made up of a few turns of large-diameter wire
    d. High resistance and made up of a many turns of small-diameter wire

    Answer _______
11. The series-type DC motor has
    a. Good speed regulation
    b. Low starting current
    c. High starting torque
    d. Both a and c

    Answer _______
12. The no-load speed of a series motor can increase to the point of damaging the motor. (True/False) ______________
13. The shunt-type DC motor is known to have
    a. High staring torque
    b. Good speed regulation
    c. Poor voltage regulation
    d. Heavy-gauge field windings

    Answer _______

14. The unmarked armature and shunt field leads for the DC shunt motor shown in Figure 5-5 are to be identified using an ohmmeter. Explain how you would proceed to identify the armature and field leads.

____

____

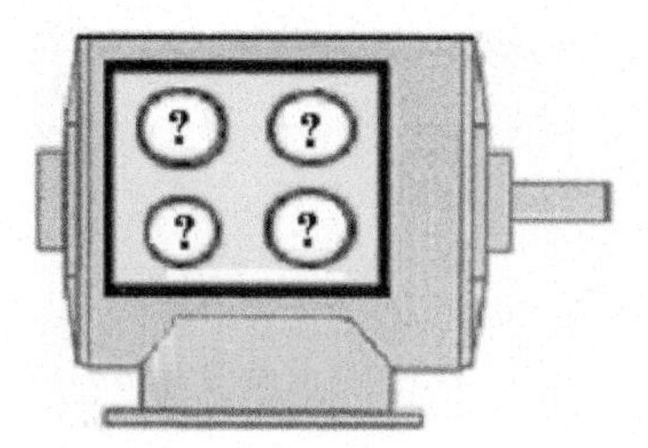

Figure 5-5 DC shunt motor for Question 14.

15. The field winding of a shunt motor is always connected across the armature and cannot be separately excited or connected to a voltage source other than that of the armature. (True/False) ______________

16. Label the field and armature leads for the compound generator shown in Figure 5-6.

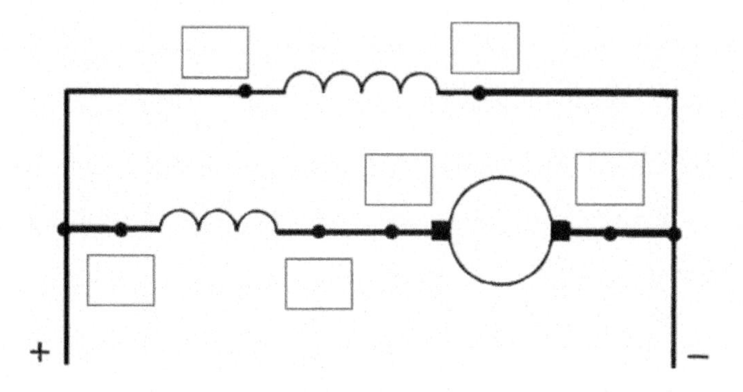

Figure 5-6 Compound generator for Question 16.

17. Identify the type of wound-field DC motor represented by each of the speed characteristic curves shown in Figure 5-7.

a. _______________

b. _______________

c. _______________

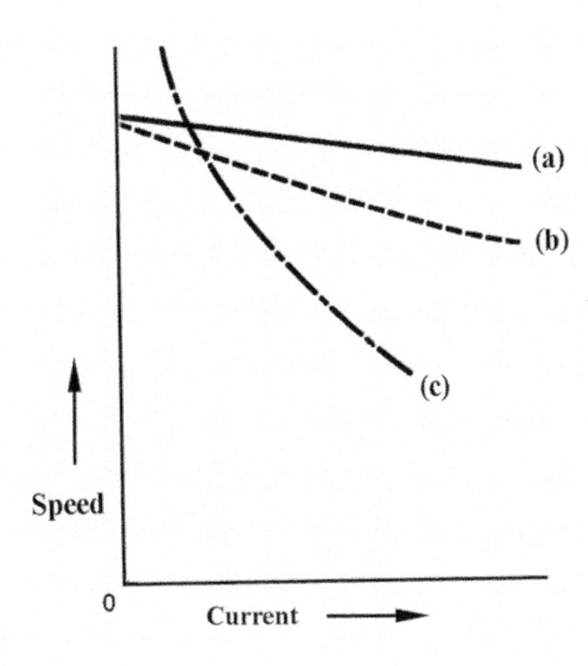

Figure 5-7 Characteristic curves for Question 17.

18. A DC motor is to be used to operate the hoist motor of an overhead crane. Which of the following DC motors would be best suited for this application?

a. Permanent magnet

b. Series

c. Shunt

d. Compound

Answer _______

19. The speed and torque characteristics shown in Figure 5-8 are that of a

_______________ DC motor.

a. Permanent magnet

b. Series

c. Shunt

d. Compound

Answer _______

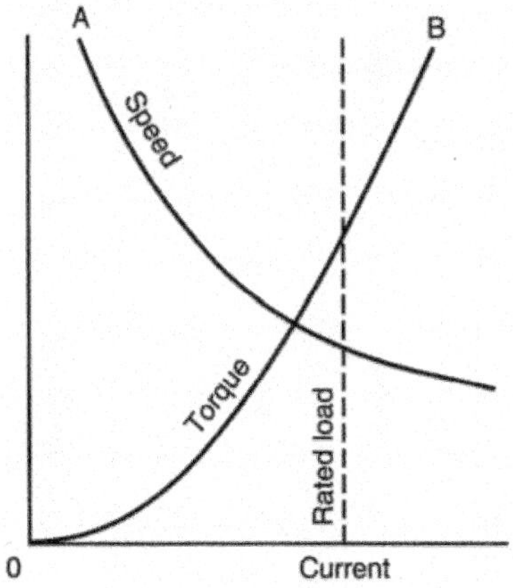

Figure 5-8 Speed and torque characteristics for Question 19.

20. The direction of rotation of a shunt-wound DC motor is changed by

a. Reversing the current flow through the armature

b. Reversing the leads of the DC voltage source

c. Reversing the current flow through the field

d. Either a or c

Answer _______

21. The voltage induced into the armature of a DC motor is known as the

a. Armature reaction

b. *IR* voltage drop

c. Applied voltage

d. CEMF

Answer _______

22. The armature current of a 125-V DC motor is measured and found to be 5 A. Determine the amount of counter EMF being generated in the armature if the armature resistance is 1.5 Ω.

Answer _______

23. At the moment of starting, no CEMF is generated in the armature of a DC motor. (True/False) _______________

24. Armature reaction in a DC motor causes
    a. A shift in the neutral plane
    b. Reduction of motor torque
    c. Arcing at the brushes
    d. All of these

    Answer _______

25. Stator poles placed between main field poles, as shown in Figure 5-9, are called _______________ and are used to minimize the effects of armature _______________.

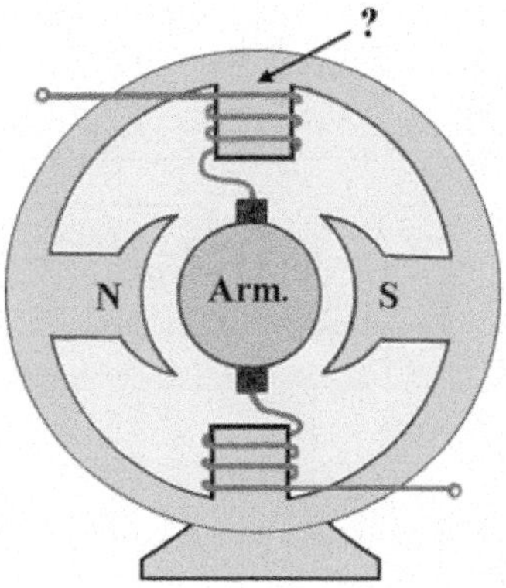

Figure 5-9 Stator poles for Question 25.

26. Determine the full-load speed for a motor with a no-load speed of 2,500 rpm and 2.5 percent speed regulation.

    Answer _______

27. The _______________ speed of a DC motor is the speed at which the motor will operate with full rated armature voltage and field current applied.

28. Increasing the voltage applied to the armature of a DC motor will cause it to run slower. (True/False) _______________

29. Decreasing the shunt field current of a DC motor will cause it to run slower. (True/False) _______________

30. Armature-controlled DC motors are capable of providing rated torque at any speed between zero and the base speed of the motor. (True/False) _______________

31. If power to the field circuit of a shunt motor is lost during operation, the motor can accelerate to a dangerously high speed. (True/False) _______________

32. Shunt motors are normally started with weakened field current. (True/False) _______________

33. Answer each of the following with reference to the DC motor drive block diagram shown in Figure 5-10.

    a. The two basic sections of the drive are the _______________ section and the _______________ section.

    b. What signal sets the desired speed?

    ____

    c. What might the feedback device be?

    ____

    d. For the block diagram of Figure 5-10, what is the function of the converter block? ____

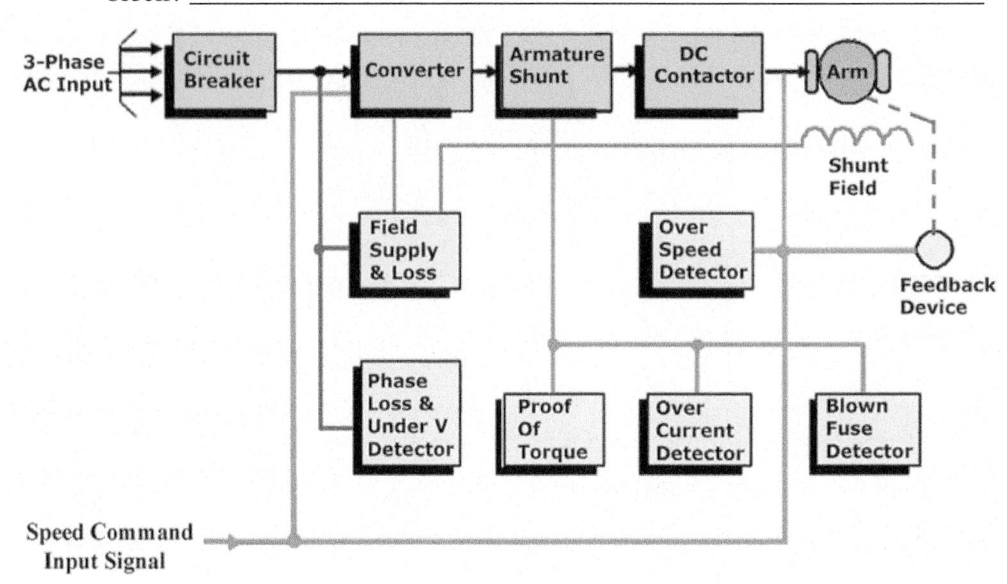

Figure 5-10 DC motor drive for Question 33.

34. With brushless type DC motors current is delivered through electronic switches into the coils on the rotor. (True/False) _______________

35. The stator coil windings of a DC brushless motor are sequentially energized. (True/False) _______________

36. The rotor of a brushless DC motor is a

    a. Wye connected coil configuration
    b. Delta connected coil configuration
    c. permanent magnet
    d. electromagnet

37. The type of brushless DC motor sensors shown are _______________ sensors.

    a. photoelectric
    b. proximity
    c. pressure
    d. Hall effect

38. Brushless DC motors are known for their ease of maintenance due to a lack of a mechanical _______________.

PART 3 Quiz: Three-Phase Alternating Current Motors

1. A _______________ magnetic field is the key to the operation of AC motors.
2. The synchronous speed of an AC induction motor is always greater than the actual rotor speed. (True/False) _______________
3. The higher the frequency of the power source, the _______________ (faster/slower) the speed of an AC induction motor.
4. The higher the number of stator poles, the _______________ (faster/slower) the speed of an AC induction motor.
5. The synchronous speed of an eight-pole, 60-Hz, AC induction motor would be _______________.
6. Figure 5-11 shows a _____________________-type rotor.

Figure 5-11 Rotor for Question 6.

7. There are no slip rings or any voltage supplied to the rotor of a squirrel-cage induction motor. (True/False) _______________

8. The resistance of the squirrel-cage rotor has little effect on the operation of the motor. (True/False) _______________

9. When a load is applied to a three-phase squirrel-cage motor the rotor speed will

   a. Increase

   b. Decrease slightly

   c. Remain the same

   d. Fluctuate

   Answer _______

10. For the motor shown in Figure 5-12, when the reverse contacts close, L1, L2, and L3 are connected to _______________, _______________, and _______________ respectively.

11. If one phase to a three-phase motor is lost during operation, the motor will immediately stop on its own. (True/False) _______________

12. An induction-type motor needs slip to provide the torque for the motor. (True/False) _______________

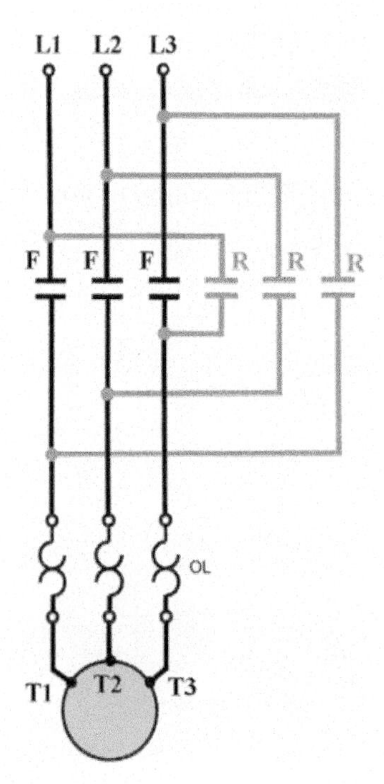

Figure 5-12

Motor circuit for Question 10.

13. Determine the slip in rpm for a four-pole induction motor with a rotor speed of 1,750 rpm, operating from a 60-Hz supply.

    Answer _______

14. Loading of an induction motor is similar to that of a transformer in that the operation of both involves changing _______________ linkages.

15. Power factor (PF) is the ratio of the _______________ to the _______________ and is a measure of how effectively the current drawn by a motor is converted into useful work.

16. The power factor of an induction motor improves when operating under load. (True/False) _______________

17. The locked-rotor starting current of an induction motor is
    a. Equal to the full-load current
    b. Less than the full-load current
    c. One-half the full-load current
    d. Several times the full-load current

    Answer _______

18. _______________-pole motors are single-winding motors that will operate at two different speeds, depending on how the windings are connected to form a different number of magnetic poles.

19. The low speed on a separate winding two-speed motor is always one-half of the higher speed. (True/False) _______________

20. Connect the dual-speed, six-lead motor shown in Figure 5-13 according to the NEMA nomenclature for high-speed operation.

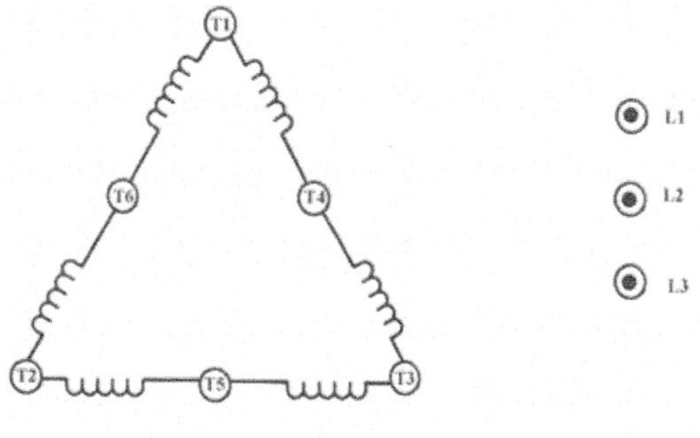

NEMA NOMENCLATURE—6 LEADS

CONSTANT TORQUE CONNECTION
Low-speed horsepower is half of high-speed horsepower.

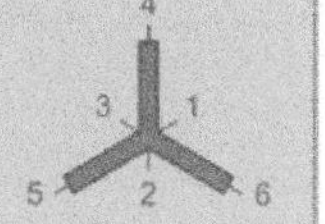

SPEED	L1	L2	L3		TYPICAL CONNECTION
HIGH	6	4	5	1&2&3 JOIN	2 WYE
LOW	1	2	3	4-5-6 OPEN	1 DELTA

Figure 5-13

Circuit for Question 20.

21. Connect the dual-voltage, nine-lead Wye-connected motor shown in Figure 5-14 according to the NEMA nomenclature for low-voltage operation.

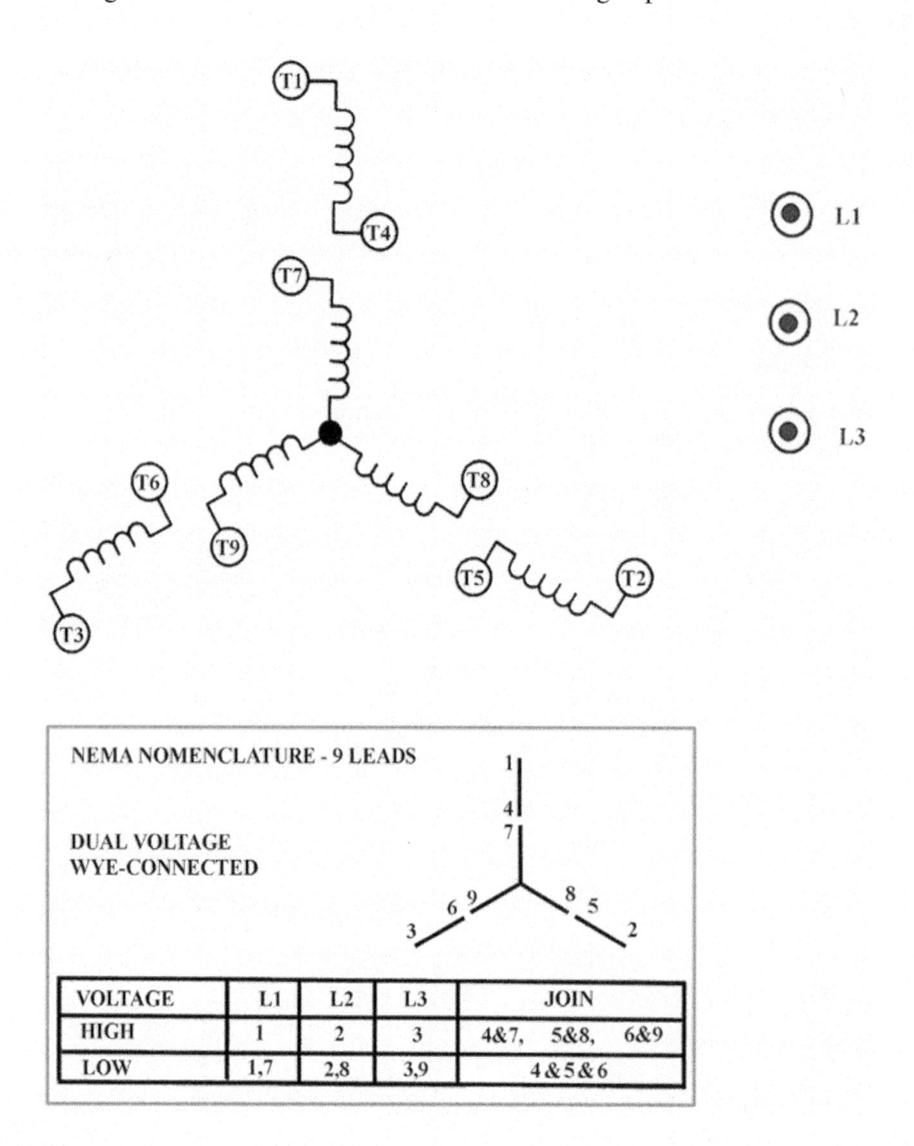

VOLTAGE	L1	L2	L3	JOIN
HIGH	1	2	3	4&7, 5&8, 6&9
LOW	1,7	2,8	3,9	4&5&6

Figure 5-14 Wye-connected motor for Question 21.

22. Answer each of the following with reference to the wound-rotor induction motor circuit shown in Figure 5-15.
    a. The motor is normally started with ______________ (full/zero) external resistance in the rotor circuit that is gradually reduced to ______________ (full/zero).
    b. The motor produces a very ______________ (high/low) starting torque from zero speed to full speed at a relatively ______________ (high/low) starting current.
    c. The direction of rotation is changed by interchanging any two of the external resistor leads. (True/False) ______________

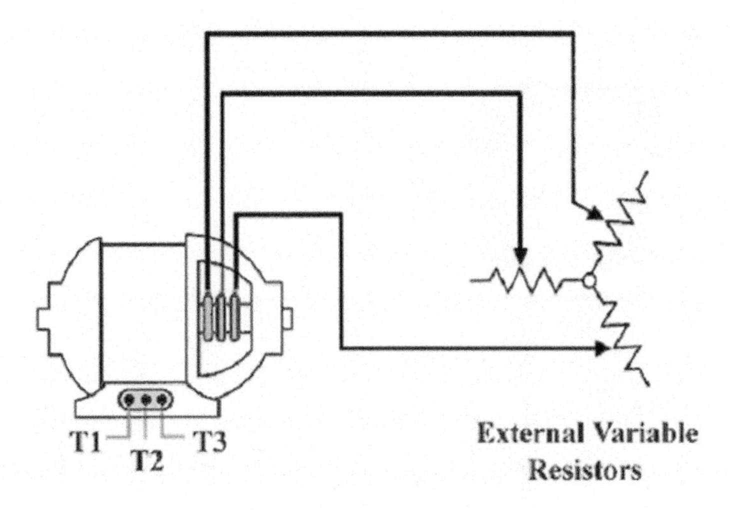

Figure 5-15 Induction motor circuit for Question 22.

23. A three-phase synchronous motor requires
    a. Both AC and DC voltages
    b. Only an AC voltage
    c. Only a DC voltage
    d. Two levels of AC voltage

    Answer _______
24. The three-phase synchronous motor has zero slip. (True/False) ______________

25. For the three-phase distribution system shown in Figure 5-16, state whether the system power factor meter is most likely to move in a lead or lag direction for each of the following operating changes:

a. Decrease in the number of inductive motor loads connected to the system _______________

b. Full-load conditions on all inductive-type motor loads connected to the system _______________

c. Increase in the synchronous motor rotor field current _______________

d. Decrease in the synchronous motor rotor field current _______________

e. Connection of a three-phase furnace resistive heater element load to the system _______________

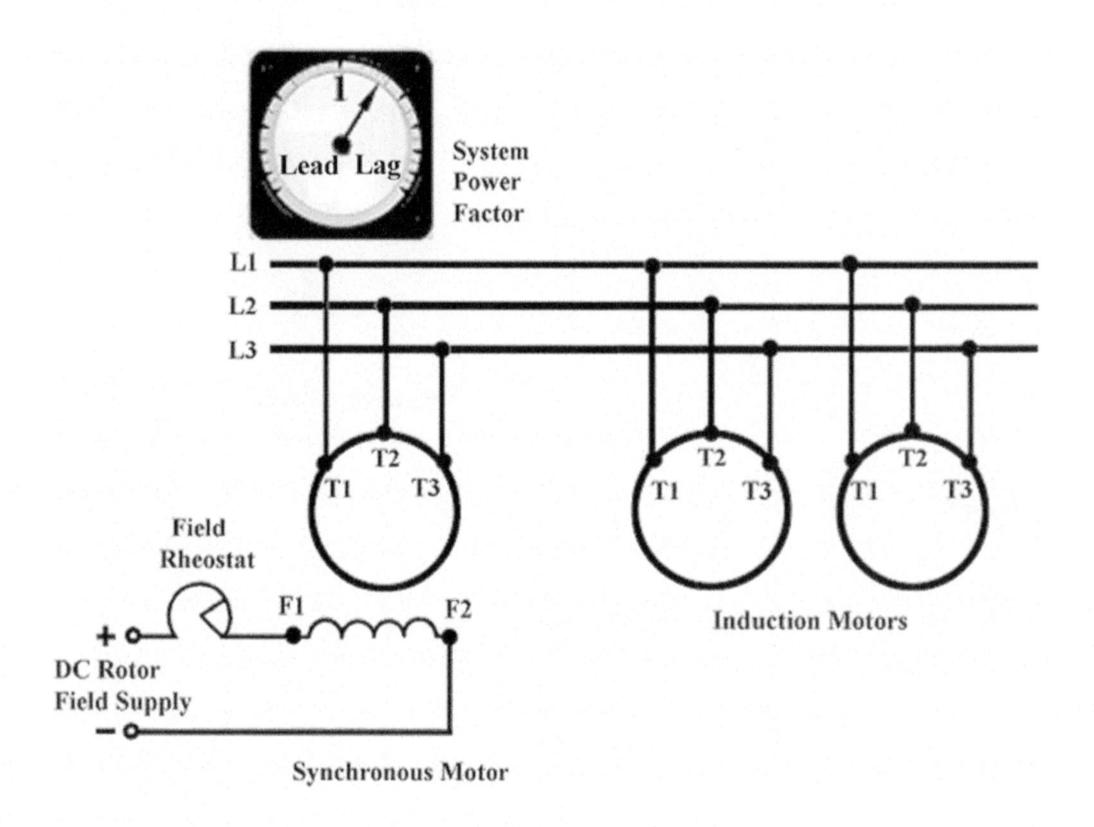

Figure 5-16 Three-phase distribution system for Question 25.

PART 4 Quiz: Single-Phase Alternating Current Motors

1. Single-phase induction motors are used in residential and commercial applications where _______________ power is not available.
2. Unlike three-phase induction motors, single-phase types are not self-starting. (True/False) _______________
3. Single-phase motors are classified by their _______________ and _______________ characteristics.
4. Identify the major components of the split-phase motor shown in Figure 5-17.
   a. ____
   b. ____
   c. ____
   d. ____

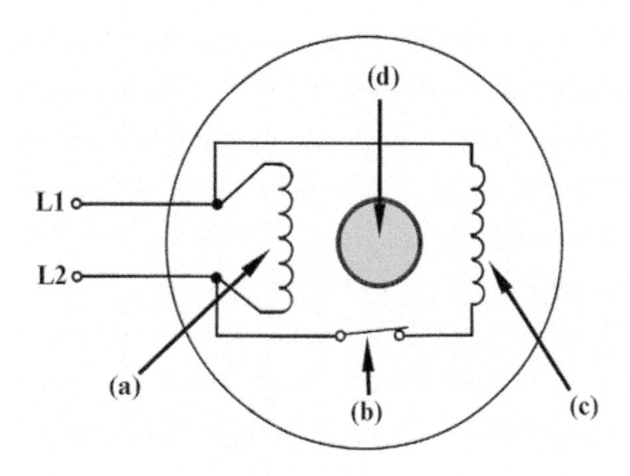

Figure 5-17 Split-phase motor for Question 4.

5. The current flow through the start and run windings of a split-phase motor is in phase with each other. (True/False) _______________
6. The direction of rotation on a split-phase motor can be reversed by interchanging the connections to
   a. The run winding
   b. The start winding
   c. Either a or b
   d. Both a and b

   Answer _______

7. Connect the dual-voltage split-phase motor shown in Figure 5-18 for 230-V CCW operation.

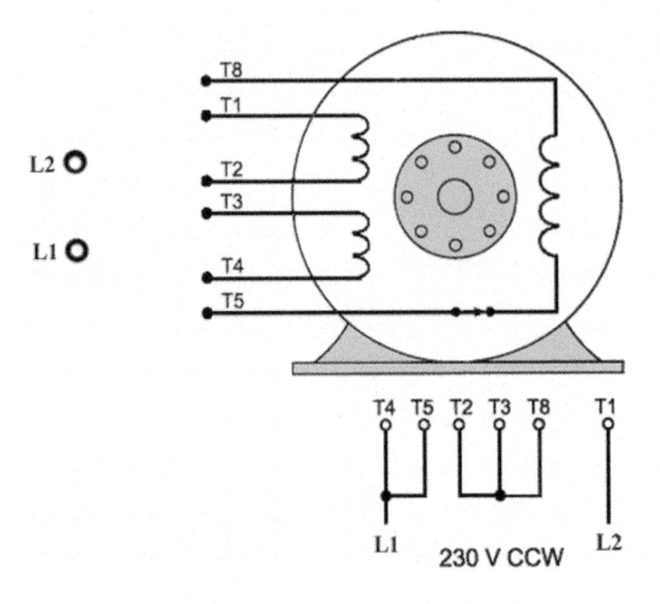

Figure 5-18 Dual-voltage, split-phase motor for Question 7.

8. Compared to split-phase motors, capacitor-start motors provide increase starting _______________ with less starting _______________.

9. Complete the internal wiring of the capacitor-start motor shown in Figure 5-19 to the supply lines.

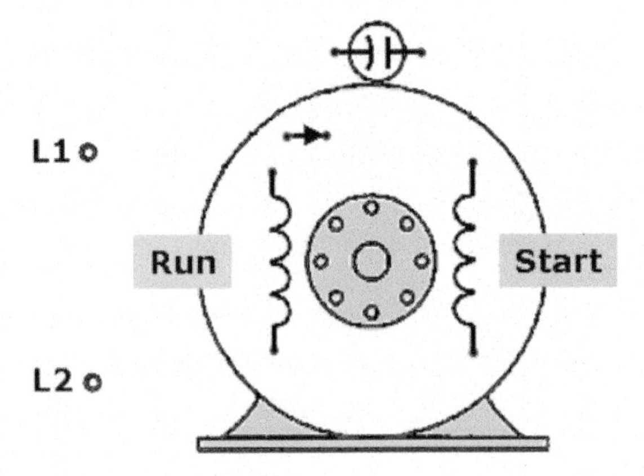

Figure 5-19 Capacitor-start motor for Question 9.

10. With capacitor-type motors, the capacitor can be a source of trouble if it becomes _______________ or _______________-circuited.

11. A permanent-capacitor motor
    a. Does not require a centrifugal switch
    b. Uses identical run and auxiliary windings
    c. Provides improvement of motor power factor
    d. All of these

    Answer _______

12. Complete the wiring for the reversing permanent-capacitor motor circuit in Figure 5-20.

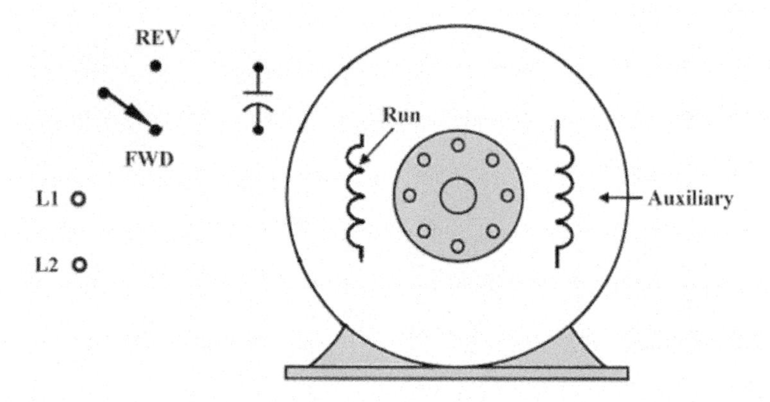

Figure 5-20 Reversing motor circuit for Question 12.

13. Capacitor-start/capacitor-run motors use two identical capacitors. (True/False) _______________

14. The capacitor-start/capacitor-run motor has the lowest starting torque of all the capacitor type motors. (True/False) _______________

15. Complete the internal wiring of the capacitor-start/capacitor run motor shown in Figure 5-21 to the supply lines.

Start Capacitor
Start Switch
Run Capacitor
L1
Run Winding
L2
Auxiliary Winding

Figure 5-21 Motor for Question 15.

16. Identify the major components of the shaded-pole motor shown in Figure 5-22.

    a. ______________________________

    b. ______________________________

    c. ______________________________

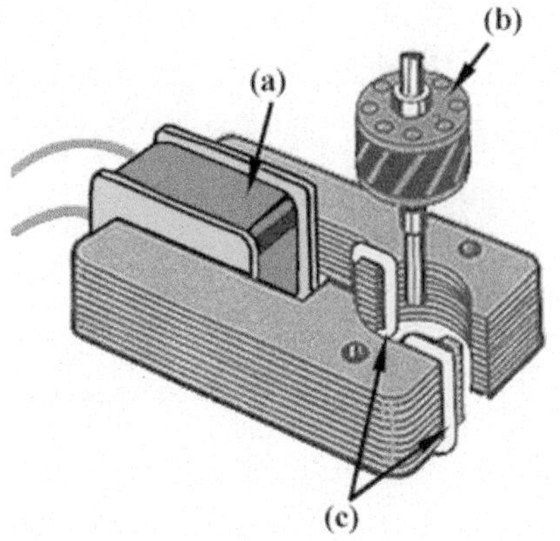

Figure 5-22

Shaded-pole motor

for Question 16.

17. Shaded-pole motors are used to drive

    a. Compressors

    b. Woodworking machinery

    c. Small fans

    d. All of these

    Answer _______

18. Identify the major components of the universal motor shown in Figure 5-23.

    a. ______________________________

    b. ______________________________

    c. ______________________________

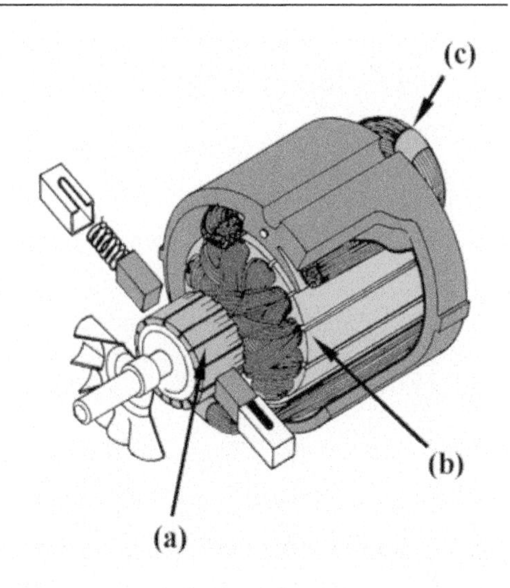

Figure 5-23

Universal motor

for Question 18.

19. A universal motor is designed to operate from
    a. An AC single-phase supply
    b. A three-phase supply
    c. A DC supply
    d. Both a and c

    Answer _______

20. Complete the internal wiring of the universal motor shown in Figure 5-24 to the supply lines.

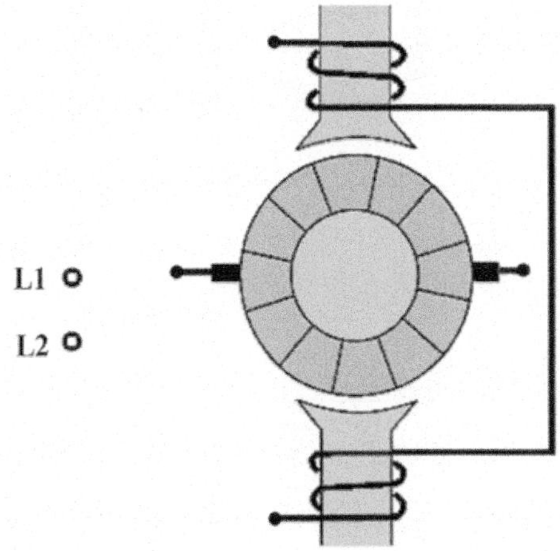

Figure 5-24

Universal motor for Question 20.

21. Reversing the direction of rotation of a universal motor is accomplished by interchanging the two line leads. (True/False) _____________

PART 5 Quiz: Alternating Current Motor Drives

1. A variable-frequency drive is a system for controlling the rotational speed of an AC motor by controlling the _____________ of the electrical power supplied to the motor.
2. The higher the frequency of the power supplied to an AC motor, the _____________ (faster/slower) the motor will turn.

3. Identify the major sections of the variable-frequency drive shown in Figure 5-25.

   a. ______________________________

   b. ______________________________

   c. ______________________________

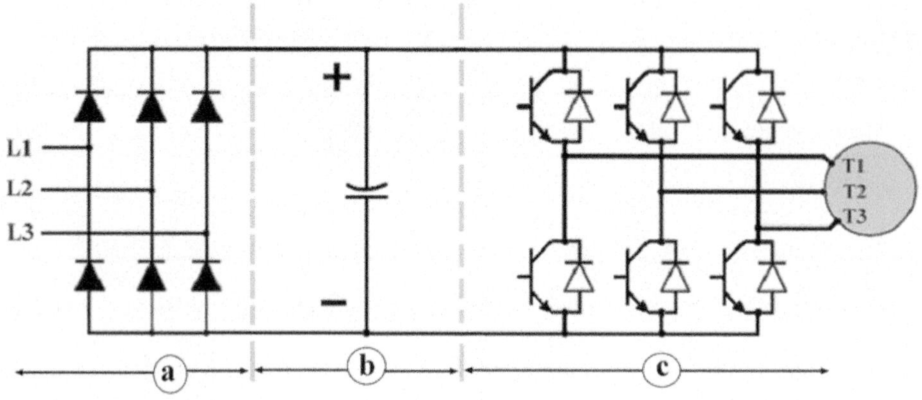

Figure 5-25 Variable-frequency drive for Question 3.

4. The inverter section of a variable-frequency drive rectifies the incoming AC power and converts it to DC. (True/False) ____________

5. If an AC motor is designed to operate at 460 V at 60 Hz, the applied voltage must be reduced to ____________ V when the frequency is reduced to 15 Hz.

6. The motor voltage waveform shown in Figure 5-26 is an example of ____________ voltage control.

   a. Pulse-width modulated
   b. Direct current
   c. Sine wave
   d. Indirect

   Answer ______

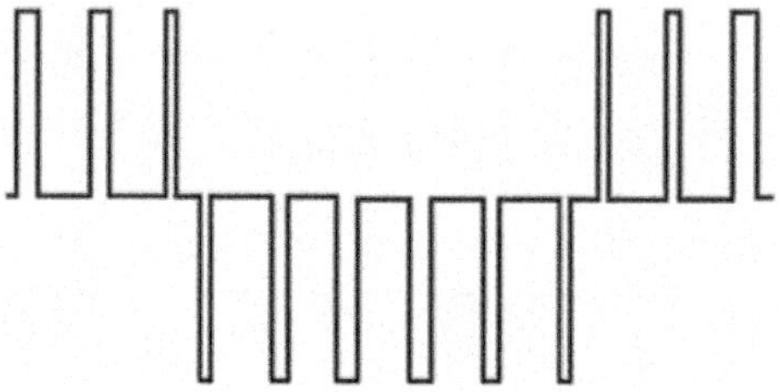

Figure 5-26 Voltage waveform for Question 6.

7. ____________ duty describes a class of motors specifically designed for use with variable-frequency drives.

   a. Heavy
   b. Light
   c. Inverter
   d. Constant

   Answer ______

PART 6 Quiz: Motor Selection

1. Slower motors are generally smaller, lighter, and less expensive than faster motors of equivalent horsepower rating. (True/False) _______________
2. Motors that are not fully loaded draw _______________ (more/less) current than rated nameplate current.
3. Calculate the maximum locked-rotor current for the following three-phase squirrel-cage motor: NEMA Code letter C (4.0 kVA/hp), 460 V, 100 hp.
   Answer _______
4. The NEMA _______________ letter denotes the motor's performance characteristics relating to torque, starting current, and slip.
5. Motor efficiency is the ratio of _______________ power output to the _______________ power input, usually expressed as a percentage.
6. The core, copper, and mechanical losses associated with the operation of a motor are dissipated as _______________ through the body of the motor.
7. Frame sizes are used to ensure the interchangeability of motors having the same frame size. (True/False) _______________
8. If a motor with a nameplate rating of 50 Hz is operated from a 60-Hz power supply, the speed of the motor would
   a. Increase slightly
   b. Decrease slightly
   c. Increase by about 20 percent
   d. Decrease by about 20 percent

   Answer _______
9. A constant horsepower load requires high torque at low speeds and low torque at high speeds. (True/False) _______________
10. A _______________ is an example of a constant torque load.
    a. Conveyor
    b. Fan
    c. Pump
    d. Lathe

    Answer _______

11. The standard motor ambient temperature rating is

    a. 64°F

    b. 75°F

    c. 85°F

    d. 104°F

    Answer _______

12. What is the safe maximum operating temperature for a motor rated for an ambient temperature of 40°C, temperature rise of 105°C, and a hot spot allowance of 10°C?

    Answer _______

13. Continuous-duty-cycle motors are labeled INTER on the nameplate. (True/False) _______________

14. The amount of torque produced by a motor when it is initially energized at full voltage is known as the

    a. Full-load torque

    b. Locked-rotor torque

    c. Breakdown torque

    d. Pull-up torque

    Answer _______

15. An _______________ motor enclosure has ventilating openings that permit passage of external air over and around the motor windings.

16. Hazardous location motors are designed for environments where explosive or ignitable _______________ or _______________ are present or may become present.

17. Metric motors are rated in _______________ rather than horsepower.

18. Which of the following symbols represents a European motor standard?

    a. UL

    b. CSA

    c. CE

    d. UCAL

    Answer _______

PART 7 Quiz: Motor Installation

1. Which of the following types of motor mountings will allow direct mounting with a pump?
   a. Adjustable base
   b. Rigid base
   c. Resilient base
   d. NEMA C face mount

   Answer _______
2. Moving a motor or placing a shim under the feet of the motor is not normally part of the alignment process. (True/False) ______________
3. Determine the speed of the load pulley for the belt drive system shown in Figure 5-27.

   Answer _______

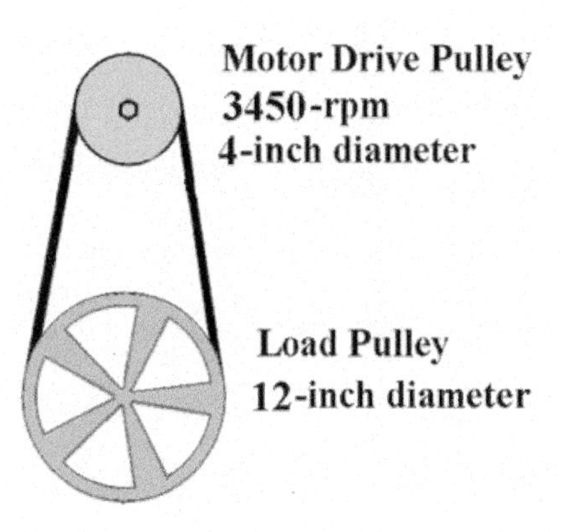

Figure 5-27 Belt drive system for Question 3.

4. Figure 5-28 illustrates an example of ______________ misalignment.

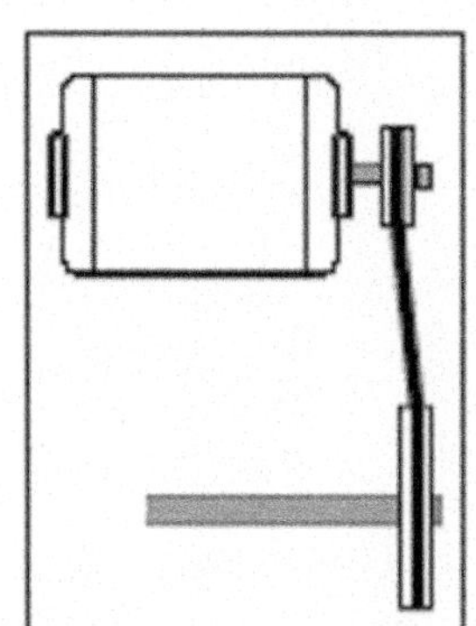

Figure 5-28 Belt drive for Question 4.

5. Motors come equipped with _______________ properly lubricated to prevent metal-to-metal contact of the motor shaft.
6. Article _______________ of the National Electrical code (NEC) contains recommendations covering motors and motor controls.
7. The equipment grounding conductor for motor circuits may be
   a. An insulated conductor run with the circuit conductors
   b. A bare conductor run with the circuit conductors
   c. The metal raceway containing the circuit conductors
   d. All of these

   Answer _______
8. The color _______________ is reserved for an insulated grounding conductor.
9. Bearing current is the result of induced voltage in the motor rotor seeking a path to ground through the motor bearings. (True/False) _______________
10. Given: a single 10-hp, three-phase, 208-V squirrel-cage motor. According to the NEC, determine each of the following:
    a. Full-load current (FLC) of the motor _______________
    b. Minimum ampacity of the branch circuit conductors _______________
    c. Branch circuit wire size for type THWN CU conductors _______________
11. For the three-phase voltage supply illustrated in Figure 5-29, determine each of the following:
    a. Average voltage _______________
    b. Maximum deviation from the average voltage _______________
    c. Percent voltage unbalance _______________

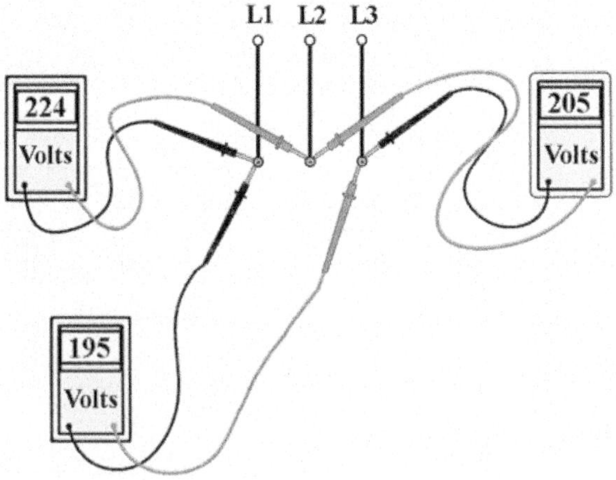

Figure 5-29 Three-phase supply for Question 11.

12. Unbalanced motor voltages applied to a polyphase induction motor may cause unbalanced currents resulting in motor overheating. (True/False)

    ____

13. What type of built-in thermal motor protection would most likely be used for roof ventilation fan motor application?

    a. Normally open type

    b. Totally enclosed type

    c. Automatic reset type

    d. Manual reset type

    Answer _______

PART 8 Quiz: Motor Maintenance and Troubleshooting

1. Two things unique to DC motor maintenance are _______________ and _______________ care.

2. The typical minimum motor insulation resistance for a motor rated for 600 V or less is

    a. 50 Ω

    b. 500 Ω

    c. 1.5 kΩ

    d. 1.5 MΩ

    Answer _______

3. Motors that are used continuously are prone to moisture problems. (True/False) _______________

4. Excessive starting of a motor should be avoided because high current flow during start-up generates a great amount of _______________ within the motor.

5. Identify the major parts of the motor system shown in Figure 5-30.

   a. ____________________

   b. ____________________

   c. ____________________

   d. ____________________

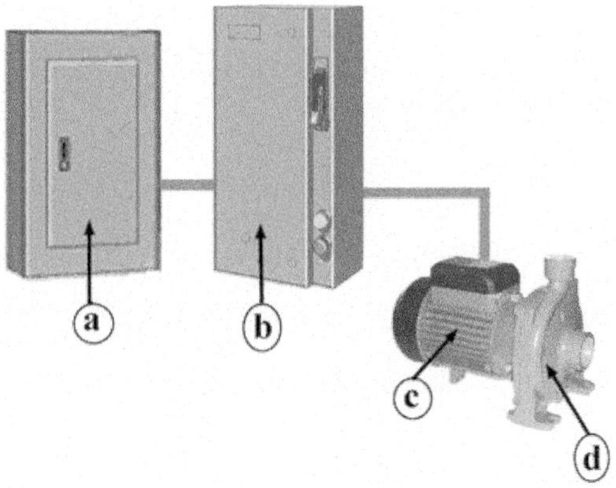

Figure 5-30
Motor system for Question 5.

6. Before beginning electrical troubleshooting of a motor, check for all possible _______________ malfunctions.

7. List six common causes for failure of any type of motor to start.

   a. ____________________

   b. ____________________

   c. ____________________

   d. ____________________

   e. ____________________

   f. ____________________

8. A bad motor bearing can cause excessive _______________ and _______________.

9. A broken or disconnected equipment _______________ conductor may cause the motor to produce an electric shock when touched.

10. A split-phase motor hums, and will run normally if started by hand. The most probable cause is

    a. A defective motor thermal overload protector

    b. The centrifugal switch is not operating properly

    c. The run winding is open

    d. The load is too high

    Answer _______

11. The start capacitor on a capacitor-start motor continuously fails. The most probable cause is
    a. The motor is being cycled on and off too frequently
    b. An open circuit in the start winding
    c. The centrifugal switch is faulted open
    d. The motor bearings are worn

    Answer _______
12. A single-phase failure of a three-phase motor is the result of an _______________ in one phase of the power supply to the motor.
13. The presence of _______________ distortion in the applied voltage to a three-phase motor will increase the motor temperature.
14. When a wound-rotor induction motor fails to start, considerable focus should be given to failed components in the rotor _______________ bank.
15. List six common causes of excessive arcing at the brushes of direct current motors.
    a. ____
    b. ____
    c. ____
    d. ____
    e. ____
    f. ____
16. Answer each of the following using the motor nameplate information shown in Figure 5-31.
    a. The motor is a _______________- phase type.
    b. The frame number _______________ gives all the critical measurements of the motor.
    c. The normal operating current of the motor when running at rated horsepower and 460 V is _______________ A.
    d. The multiplier applied to the motor horsepower is _______________.
    e. The maximum operating temperature of the motor windings is indicated by the letter _______________.
    f. The rated mechanical output of the motor is _______________.
    g. The value of the rated low voltage of the motor is _______________ V.

h. The maximum temperature of the surrounding environment that will allow rated horsepower without damage is _______________.

i. The letter that refers to the locked-rotor characteristics of the motor is _______________.

j. The rated operating frequency of the motor is _______________.

k. The input electrical watts converted to mechanical output watts or horsepower is _______________.

l. The letter that refers to the motor's operating characteristics such as the starting torque and starting current is _______________.

m. The rated speed of the motor when operated at 230 V and full load is _______________.

n. The length of time the motor can run at full load without overheating is _______________.

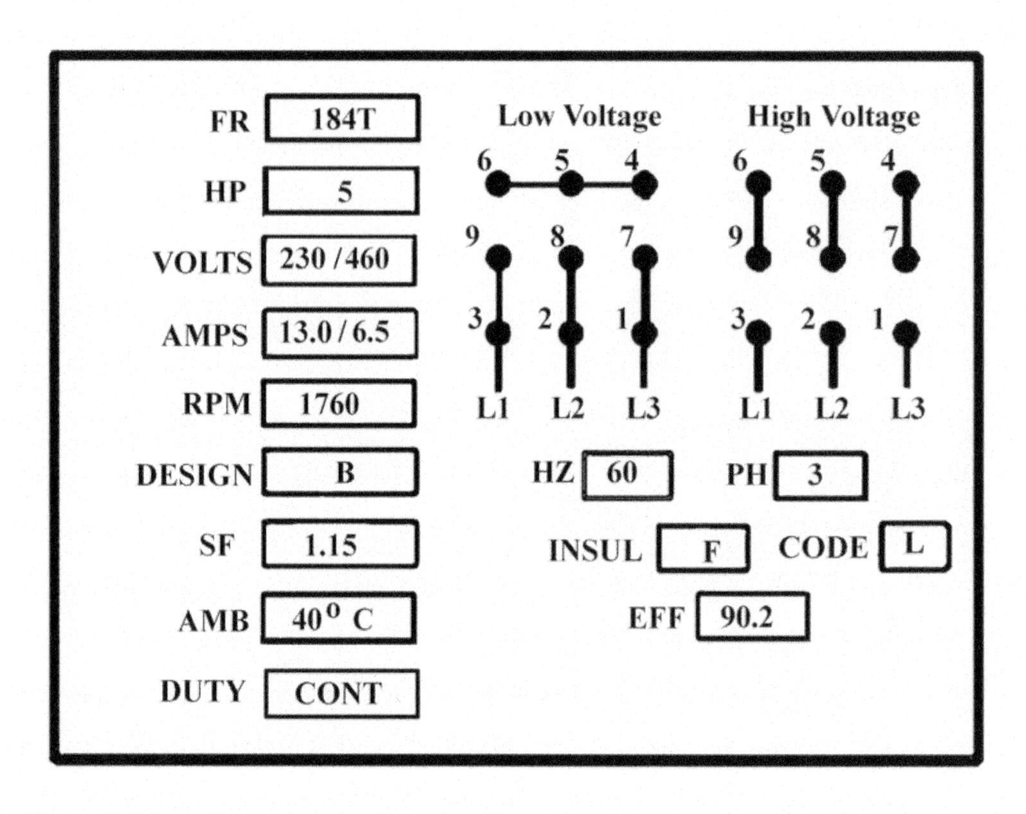

Figure 5-31 Nameplate for Question 16.

17. Which test instrument would be most likely used to measure the capacitance of a capacitor?
    a. Digital multimeter
    b. Megohmmeter
    c. Tachometer
    d. Oscilloscope

    Answer _______

18. Which test instrument would be most likely used to measure the speed of motor?
    a. Digital multimeter
    b. Megohmmeter
    c. Tachometer
    d. Oscilloscope

    Answer _______

19. Which test instrument would be most likely used to measure the insulation resistance of a motor winding?
    a. Digital multimeter
    b. Megohmmeter
    c. Tachometer
    d. Oscilloscope

    Answer _______

20. Which test instrument would be most likely used to measure the current of a motor?
    a. Clamp-on ammeter
    b. Digital multimeter
    c. Infrared thermometer
    d. Oscilloscope

    Answer _______

21. Which test instrument would be most likely used to pinpoint a hot point on the surface of a motor?
    a. Clamp-on ammeter
    b. Digital multimeter
    c. Infrared thermometer
    d. Oscilloscope

    Answer _______

22. Which test instrument would be most likely used to examine the output voltage waveform from a motor drive?

    a. Clamp-on ammeter
    b. Digital multimeter
    c. Infrared thermometer
    d. Oscilloscope

    Answer _______

23. The contacts of a centrifugal switch in a split-phase motor do not close due to a malfunction. What is the result?

    a. The motor will reach the full rated speed.
    b. Voltage will not be applied to any of the stator windings.
    c. The starting winding will burn out in a very short time.
    d. The running winding will be energized but the motor will not start.

24. A 45μF 347V motor capacitor is faulty and needs to be replaced. The original capacitor size is not available. What size capacitor is suitable to be used as the replacement?

    a. 45μF 250V
    b. 35μF 120V
    c. 55μF 347V
    d. 45μF 120V

6 Contactors and Motor Starters

PART 1 Quiz: Magnetic Contactor

Place the answers in the space provided.

1. The magnetic contactor is similar in operation to the electromechanical ______________.
2. Contactors are designed to make and break electric circuit loads of higher currents than relays. (True/False) ______________
3. Identify the components of the contactor shown in Figure 6-1.
   a. ______________________________
   b. ______________________________
   c. ______________________________

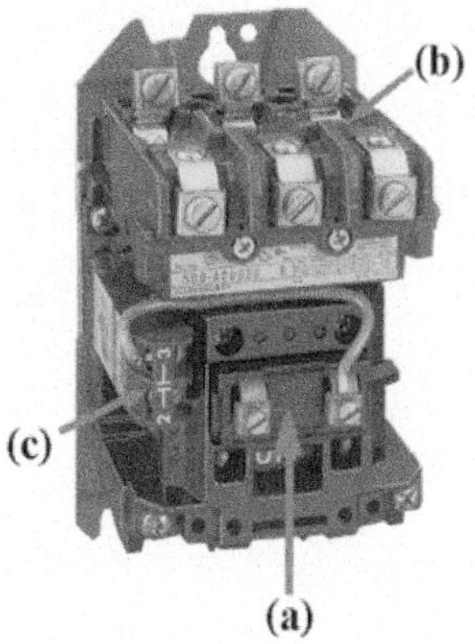

(b)
(c)
(a)

Figure 6-1 Contactor for Question 3.

4. The coil is part a magnetic contactor's ______________ circuit.
5. The contacts of a contactor are activated any time voltage is applied to the
   a. Coil
   b. Power contacts
   c. Auxiliary contacts
   d. All of these

   Answer _______
6. A ______________-pole contactor would be required to switch power to a three-phase load.

7. Identify the line side, load side, and parts of the contactor shown in Figure 6-2.

   a. ______________________________

   b. ______________________________

   c. ______________________________

   d. ______________________________

   e. ______________________________

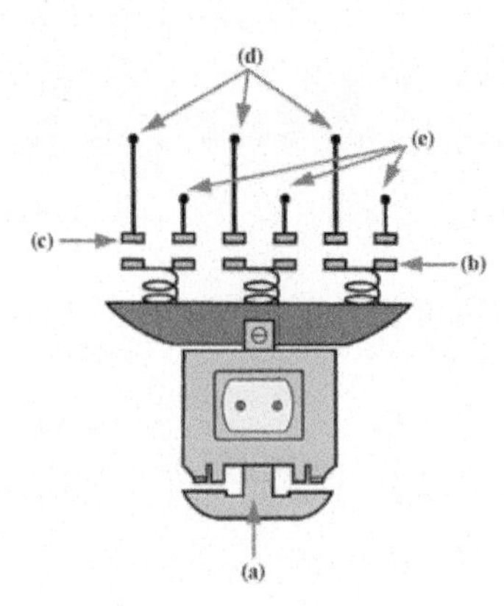

Figure 6-2 Contactor for Question 7.

8. Complete the wiring for the circuit drawn in Figure 6-3 connected so that the ON/OFF pilot switch switches the current to the three-phase heater load.

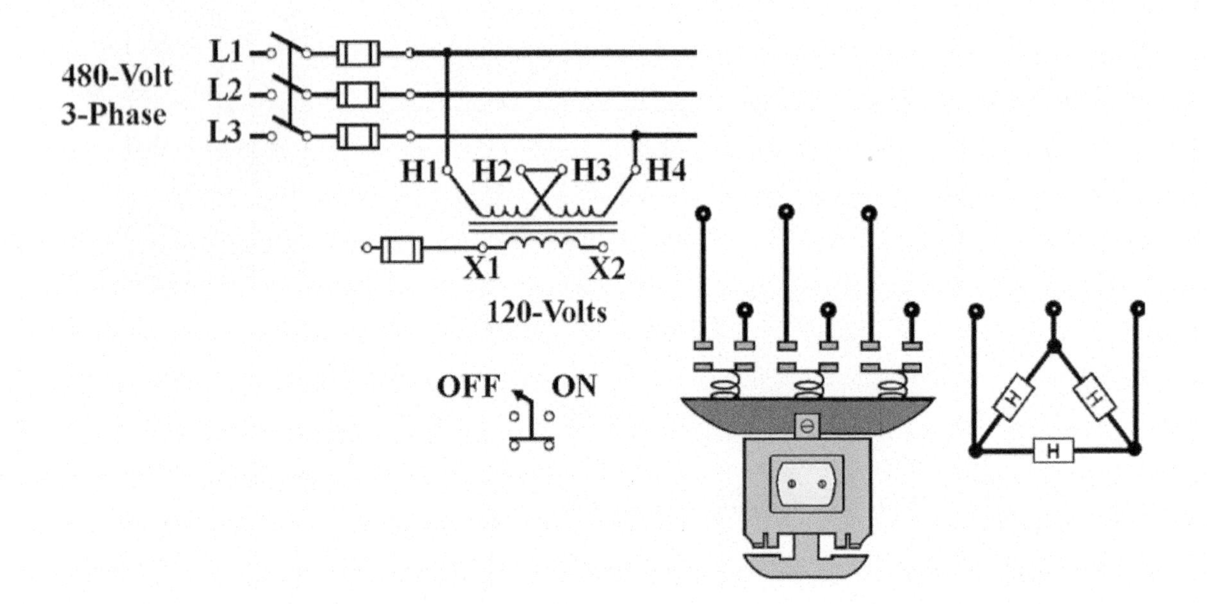

Figure 6-3 Circuit for Question 8.

9. Contactors may be used for switching motor loads when separate _______________ protection is provided.

10. If the control circuit is connected directly to the same line wires as the load, then it must be rated for a lower voltage. (True/False) _______________

11. Pilot lights operating in conjunction with a contactor are normally controlled by auxiliary contacts. (True/False) _______________

12. With an _______________ held contactor, the coil needs to be energized continuously all the time the main contacts are closed.

13. A mechanically held contactor
    a. Requires only a pulse of coil current to change state.
    b. Operates more efficiently than electrically held types.
    c. Is quieter than electrically held types.
    d. All of these

    Answer _______

14. Complete the wiring for the lighting contactor circuit drawn in Figure 6-4 with the additional ON/OFF control station wired into the circuit.

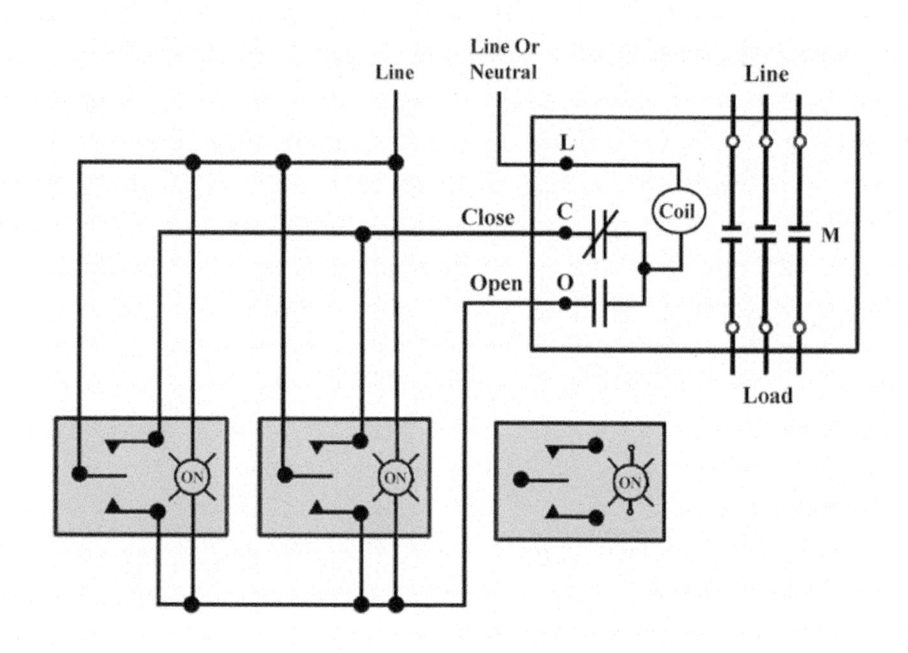

Figure 6-4 Lighting contactor circuit for Question 14.

15. The type of contactor operating mechanism shown in Figure 6-5 is the
    a. Clapper type
    b. Bell-crank type
    c. Seal-in type
    d. Horizontal-action type

    Answer _______

Figure 6-5 Contactor mechanism for Question 15.

16. A defective contactor coil will read zero or infinity, indicating a
_______________ or _______________ coil, respectively.

17. Contactor coils are designed to operate

    a. On AC only
    b. On DC only
    c. Over a range of 85 to 110 percent of rated voltage
    d. At rated voltage only

    Answer _______

18. Operating contactor coils at either lower or higher than their rated voltage may result in a high level of contact bounce. (True/False) _______________

19. The hold-in voltage rating of a contactor coil is normally

    a. The same as the pickup voltage rating
    b. Less than the pickup voltage rating
    c. Greater than the pickup voltage rating
    d. The same as the dropout voltage rating

    Answer _______

20. AC and DC contactor coils with the same voltage ratings are normally interchangeable. (True/False) _______________

21. Current flow through an AC coil is limited by both _______________ and _______________.

22. An AC contactor coil has a

    a. High ohmic resistance
    b. High inrush current
    c. A nonlaminated steel core and armature
    d. All of these

    Answer _______

23. Complete the wiring for the PLC contactor-operated circuit drawn in Figure 6-6 with an *RC* suppression module wired into the circuit.

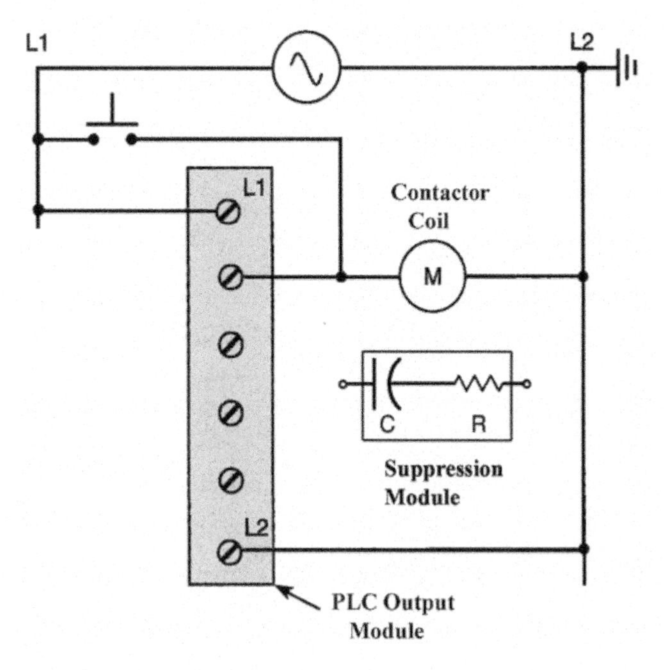

Figure 6-6 Circuit for Question 23.

24. Whenever current to a contactor coil is turned off, a high-voltage spike is generated. (True/False) _______________

25. Identify the parts of the contactor assembly shown in Figure 6-7.

    a. ________________________

    b. ________________________

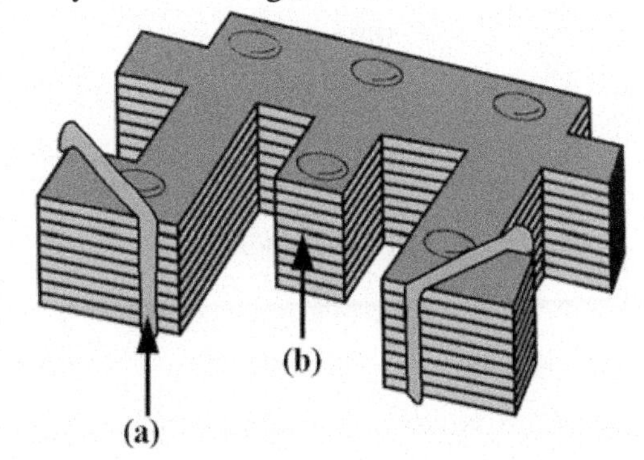

Figure 6-7 Contactor assembly for Question 25.

26. The core and armature of AC contactor assemblies are made of laminated steel to

    a. Reduce chatter

    b. Reduce the size of the contactor

    c. Reduce eddy-current flow

    d. All of these

    Answer _______

27. A broken or open shading coil will cause a contactor to become extremely noisy. (True/False) _______________

28. A contactor fails to seat properly when energized. This may result in
    a. Hum coming from the contactor
    b. Higher than normal current flow through the coil
    c. Overheating of the contactor coil
    d. All of these

    Answer _______

29. Contacts are silver-coated to reduce contact _______________.

30. Auxiliary contacts are used to switch high-current load circuits. (True/False) _______________

31. Silver contacts should never be filed. (True/False) _______________

32. Arc current flow produces additional heat, which can damage _______________ surfaces.

33. On opening under load, as the distance between contacts increases, the
    a. Resistance of the arc remains the same
    b. Resistance of the arc increases
    c. Voltage required to sustain remains the same
    d. Voltage required to sustain the arc decreases

    Answer _______

34. An AC arc is self-extinguishing, as the arc will normally extinguish as the AC cycle passes through _______________.

35. A motor would be classified as a
    a. Resistive load
    b. Nonlinear load
    c. Inductive load
    d. Capacitive load

    Answer _______

36. Identify the parts of the contactor shown in Figure 6-8.
    a. _____________________
    b. _____________________

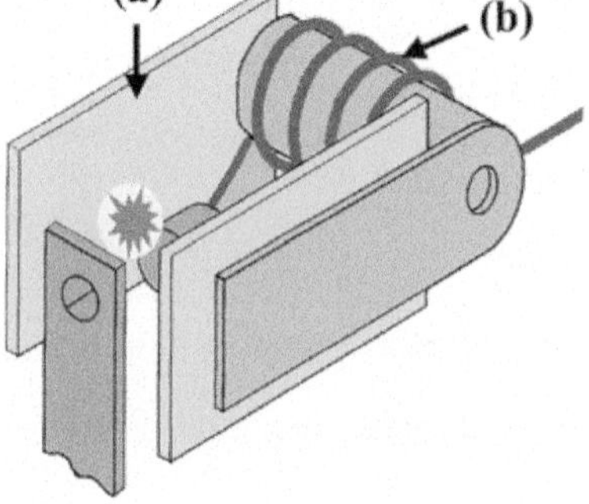

Figure 6-8 Contactor for Question 36.

37. Blowouts are connected in series with the contacts. (True/False) _______________

38. Worn contacts should always be replaced in pairs. (True/False) _______________

39. The contactor shown in Figure 6-9 would be classified as a _______________ type.

Figure 6-9 Contactor for Question 39.

40. A vacuum provides a better environment than free air for breaking the arc produced by opening contacts under load. (True/False) _______________

41. A vacuum contactor does not use a coil to operate the contacts. (True/False) _______________

42. Contactor auxiliary contacts have a current rating much higher than that of the main contacts. (True/False) _______________

43. A capacitor switching contactor uses dampening diodes across each pole designed to minimize the effects of the large charging current. (True/False) _______________

44. The main difference between contactors designed for AC power and those designed to switch DC is the enhanced ability to quench the arc so that it is not sustained any longer than necessary. (True/False) _______________

45. Capacitor switching contactors are used to switch

    a. resistive loads
    b. lighting loads
    c. motor loads
    d. power factor correction loads

PART 2 Quiz: Contactor Ratings, Enclosures, and Solid-State Types

1. The NEMA size categorizes a contactor on the basis of _______________, _______________, and _______________.

2. Which of the following NEMA size contactors would have the greatest current-carrying capacity?

    a. 0
    b. 3
    c. 6
    d. 9

    Answer _______

3. Name the contactor IEC load utilization category for each of the following:
   a. Transformer _______________
   b. Incandescent lighting installation _______________
   c. Electric furnace heater _______________
   d. Power factor correction bank _______________
4. NEMA contactors, as compared to IEC types, generally are physically downsized to provide higher ratings in a smaller package. (True/False) _______________
5. NEMA contactors are categorized by _______________, while IEC types are categorized by _______________ categories.
6. The two general classifications for NEMA contactor enclosures are _______________ location and _______________ location.
7. Which of the following NEMA-type enclosures would be selected to ensure that an internal explosion would be confined to the inside of the enclosure?
   a. Type 1
   b. Type 4X
   c. Type 7
   d. Type 12

   Answer _______
8. The IEC method for specifying the enclosures of electrical equipment is the same as that used by NEMA. (True/False) _______________
9. Solid-state switching refers to interruption of power by _______________ means.
10. A solid-state contactor
    a. Has no contacts
    b. Is silent in operation
    c. Produces no arc
    d. All of these

    Answer _______
11. Identify the leads of the SCR solid-state switch shown in Figure 6-10.
    a. _____________________
    b. _____________________
    c. _____________________

(b) (c) (a)

Figure 6-10 Solid-state switch for Question 11.

12. With reference to the SCR testing circuit shown in Figure 6-11, identify the SCR fault, if any, commonly associated with each of the following scenarios:

    a. Light turns on as soon as the power is applied.

    ____

    b. Light turns on when the ON button is pressed and remains on when the button is released.

    ____

    c. Light fails to turn on when the ON button is pressed and power is applied.

    ____

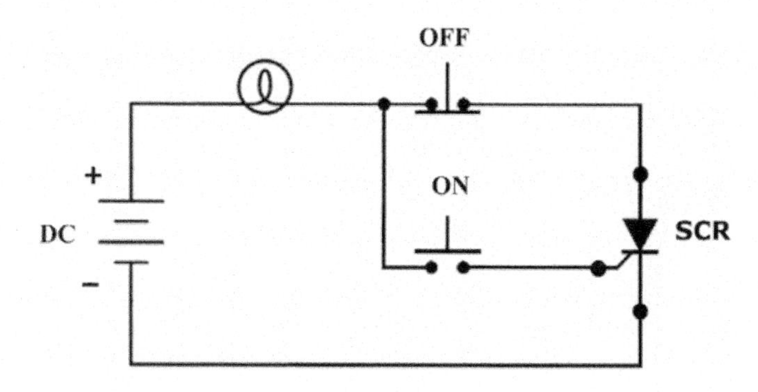

Figure 6-11 Testing circuit for Question 12.

13. Zero-fired control of an SCR refers to triggering the SCR on at the maximum point of the sine wave so that no current is being switched under load. (True/False) ______________

14. Any device that makes use of a coil of wire for its operation can be classed as a(n) ______________ load.

    a. Capacitive
    b. Resistive
    c. Passive
    d. Inductive

    Answer ________

15. Which of the following loads can create voltage spikes?
    a. Transformers
    b. Relay coils
    c. Solenoids
    d. All of these
    Answer ________
16. Unlike magnetic contactors, solid-state types can be switched ON and OFF by a lower-power digital control signal. (True/False) ______________

PART 3 Quiz: Motor Starters

1. Complete the magnetic motor starter wiring diagram in Figure 6-12 according to the control ladder diagram. Do not make any wire splices or additional terminal connections on the wiring diagram. All connections must run from terminal screw to terminal screw.

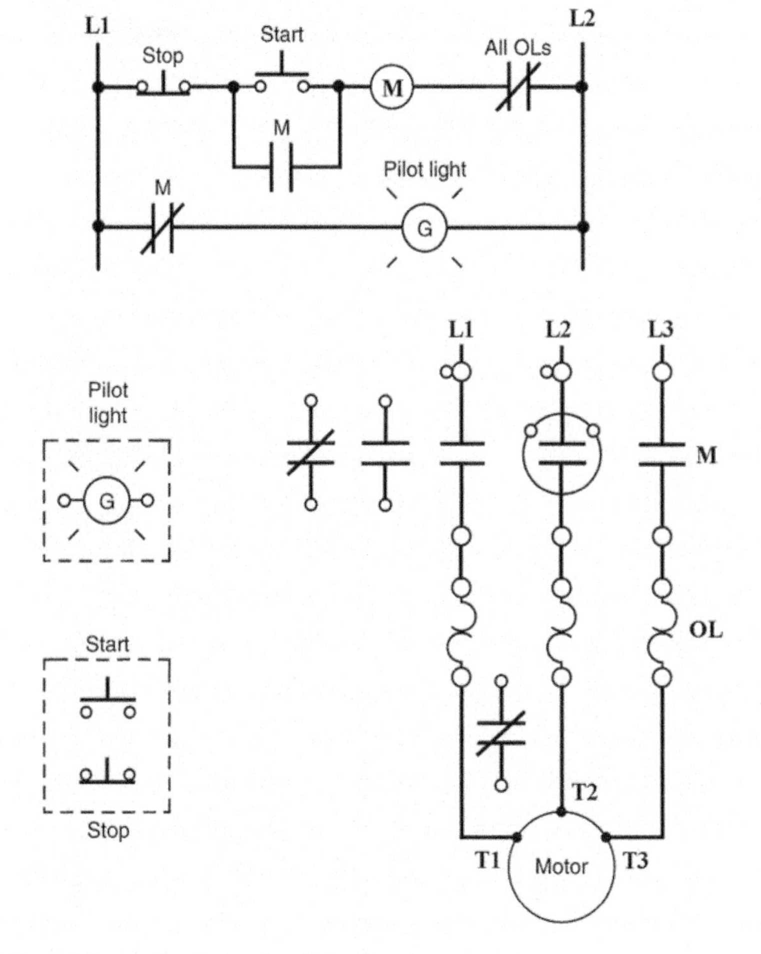

Figure 6-12 Magnetic starter for Question 1.

2. In its most common form, a motor starter is a contactor with
   a. An overload protection device
   b. An armature
   c. A circuit breaker
   d. A push button

   Answer _______

3. Enclosures protect people from electric shock. (True/False) _______________

4. Identify the major NEC requirements for the motor installation shown in Figure 6-13.
   a. ____
   b. ____
   c. ____
   d. ____
   e. ____

5. The starting current of a motor is normally several times higher than the rated full-load nameplate current. (True/False) _______________

6. A Class _______________ overload relay will trip an overloaded motor offline within 20 seconds at six times full-load amperes.
   a. 20
   b. 10
   c. 40
   d. 35

   Answer _______

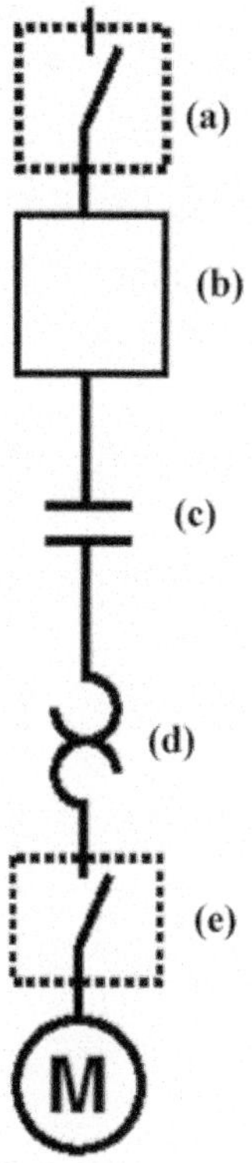

Figure 6-13 Motor installation for Question 4.

7. Once tripped, overload relays normally require a cooling-off period before they can be reset. (True/False) _______________

8. Overload relays can be classified as being _______________, _______________, or _______________.

9. External overload protection devices, which are mounted in the starter, attempt to simulate the heating and cooling of a motor by sensing the _______________ flowing to it.

10. A thermal overload relay uses a heater element connected in parallel with the motor supply. (True/False) _______________

11. A thermal overload relay protecting a three-phase motor requires _______________ heaters.

    a. 1
    b. 2
    c. 3
    d. 4

    Answer _______

12. Identify the parts of the thermal-overload relay shown in Figure 6-14.

    a. ____________________________
    b. ____________________________

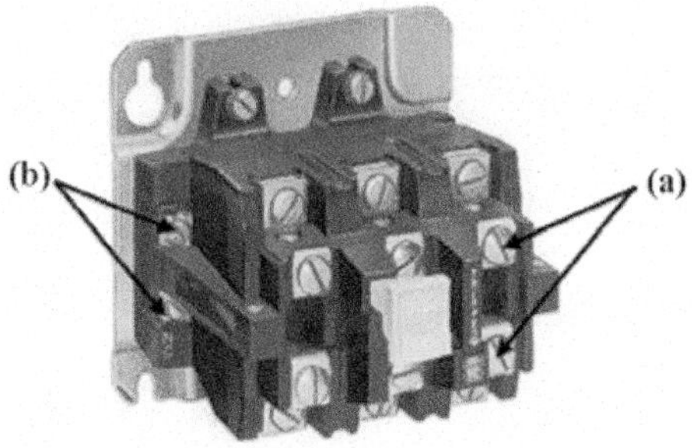

Figure 6-14 Relay for Question 12.

13. Identify the parts of the eutectic overload relay shown in Figure 6-15.

    a. ____________________________
    b. ____________________________

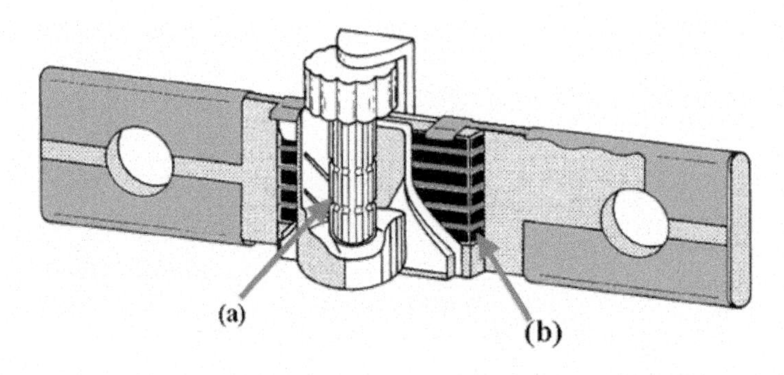

Figure 6-15 Eutectic overload relay for Question 13.

14. Identify the parts of the bimetallic overload relay shown in Figure 6-16.

   a. ____________________

   b. ____________________

   c. ____________________

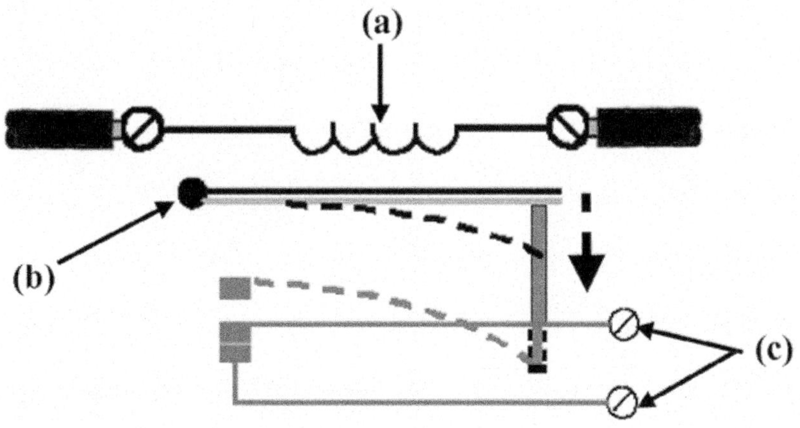

Figure 6-16 Bimetallic overload relay for Question 14.

15. Selection tables for thermal overload relays normally list OL heaters according to motor ______________.

16. Ambient temperatures will not normally affect the tripping time of a thermal overload relay. (True/False) ______________

17. Normally, in selecting OL heater sizes from a manufacturer's table, it is assumed that the motor

   a. Has a minimum service factor of 1.1

   b. Temperature rise is not over 104°F

   c. Will be protected up to 125 percent of the nameplate FLC rating

   d. All of these

   Answer _______

18. An electronic overload relay measures motor current directly through a current ______________.

19. Electronic overload relays

   a. Can be connected directly to the contactors

   b. Are easily adjustable to a wide range of full-load motor currents

   c. Reduce the amount of heat generated by the starter

   d. All of these

   Answer _______

20. For the electronic overload relay shown in Figure 6-17, identify the function or connection of each part.

a. ______________________________

____

b. ______________________________

____

c. ______________________________

____

d. ______________________________

____

e. ______________________________

f. ______________________________

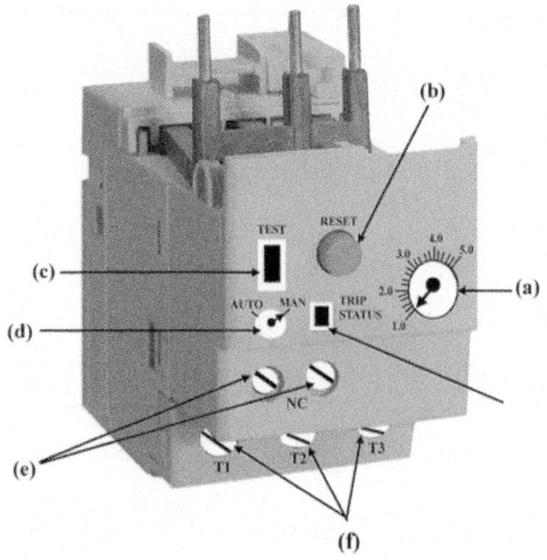

Figure 6-17 Electronic overload relay for Question 20.

21. Dual-element fuses, when properly sized, provide

   a. Short-circuit protection
   b. Ground fault protection
   c. Overload protection
   d. All of these

   Answer _______

22. Identify the elements of the dual-element fuse shown in Figure 6-18.

   a. ______________________________
   b. ______________________________

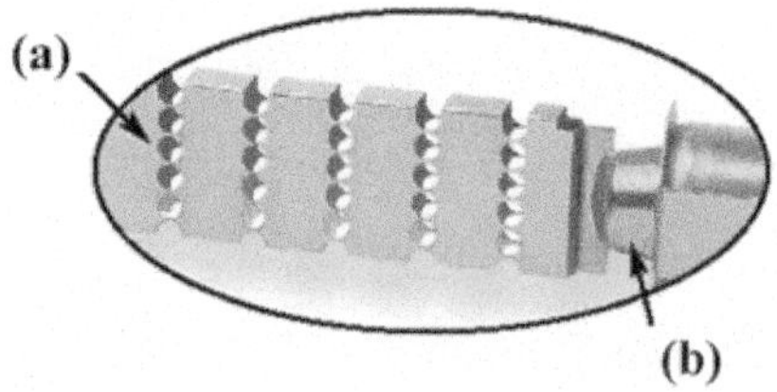

Figure 6-18 Dual-element fuse for Question 22.

23. Identify each type IEC motor control symbol shown in Figure 6-19.

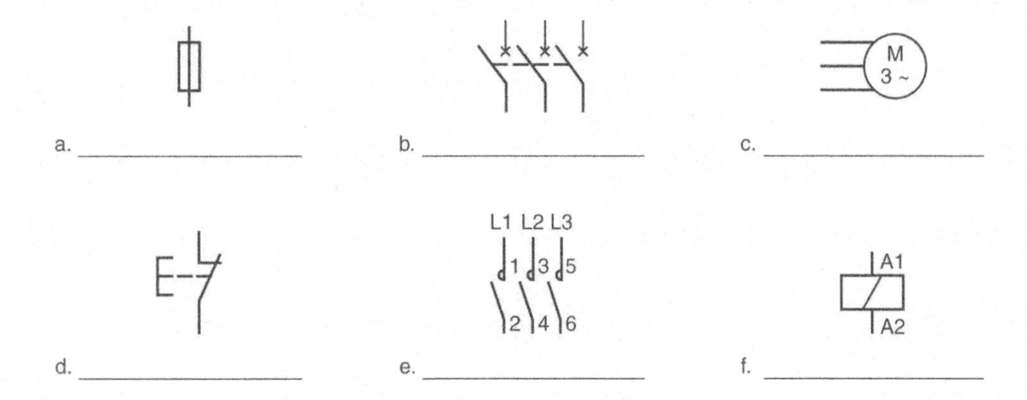

Figure 6-19 IEC symbols for Question 23.

7 Relays

PART 1 Quiz: Electromechanical Control Relays

Place the answers in the space provided.

1. The two basic parts of an electromechanical relay are the ______________ and the ______________.
2. The two circuits associated with a relay are the
   a. Primary and secondary
   b. Control and load
   c. Normally open and normally closed
   d. High and low

   Answer _______
3. Complete the relay wiring diagram for Figure 7-1 to operate so that opening and closing the switch will control current to the load through the relay contact.

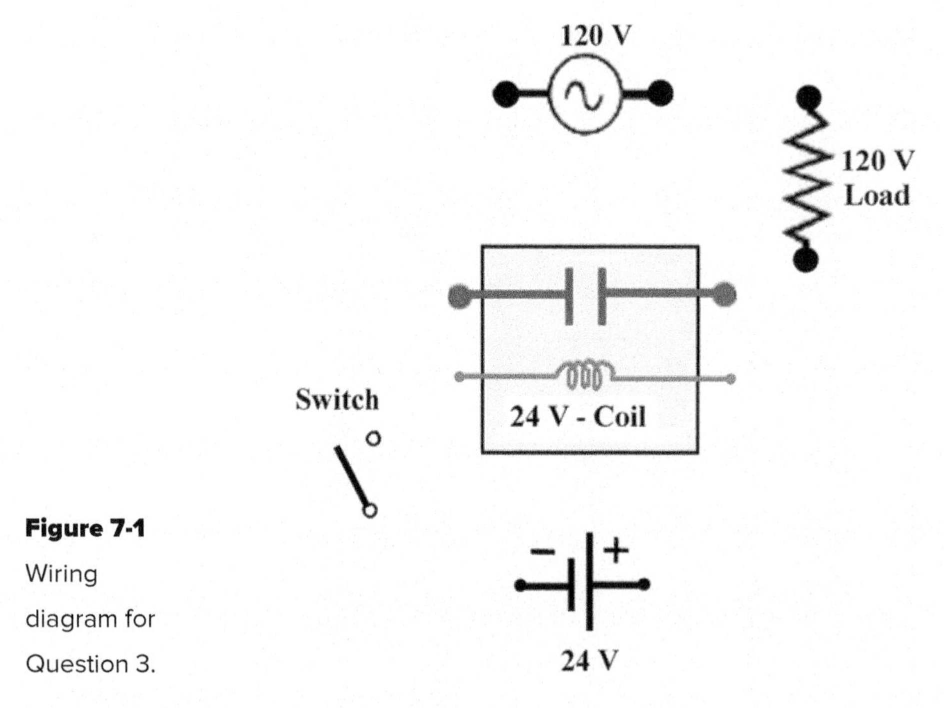

Figure 7-1 Wiring diagram for Question 3.

4. Relays are used to control large loads of 15 A or more. (True/False) ______________

5. An electromechanical relay normally will have only one ______________, but it may have a number of different ______________.
6. Normally closed contacts are defined as those contacts that are closed when the coil is ______________
7. State the correct type (N.O. or NC) and label for each of the contacts associated with the relay coil of Figure 7-2.

   a. ______________________________

   b. ______________________________

Figure 7-2 Relay coil for Question 7.

8. Which of the following is *not* a common application for a relay?
   a. To control a 120 V lighting circuit with a 12 V control circuit.
   b. To control several switching operations by a single separate current.
   c. To change alternating current to direct current.
   d. To switch a small servo motor with a low-current sensor signal.

   Answer _______
9. The coil and contacts of a relay are not normally electrically insulated from each other. (True/False) ______________
10. Complete the relay wiring diagram for Figure 7-3 to operate so that when the push button is closed, the green light switches from on to off and the red light switches from off to on.

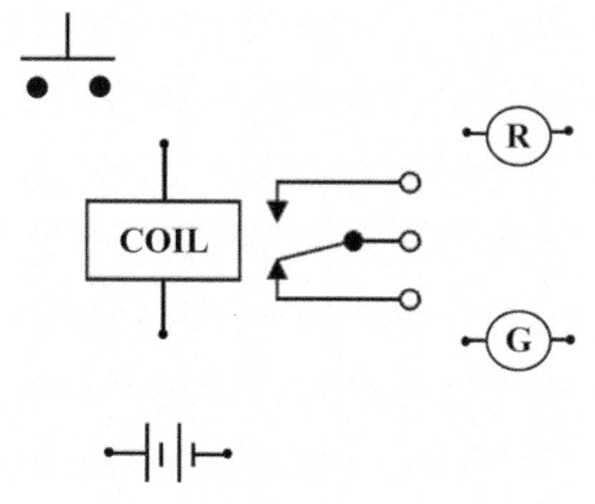

Figure 7-3 Relay wiring diagram for Question 10.

11. Relay options that aid in troubleshooting may include:

   a. ____

   b. ____

12. It is possible for a relay coil to be rated for 24 V DC and its contacts to be rated for 240 V AC. (True/False) _______________

13. Electromechanical relay contacts may be constructed as

    a. N.O. or NC contacts.

    b. Single-break or double-break contacts.

    c. Fixed and movable contacts.

    d. All of these

    Answer _______

14. State the number of poles, throws, and breaks for each of the relay contacts shown in Figure 7-4.

    a. _______________-pole

       _______________-throw

       _______________-break

    b. _______________-pole

       _______________-throw

       _______________-break

    c. _______________-pole

       _______________-throw

       _______________-break

    d. _______________-pole

       _______________-throw

       _______________-break

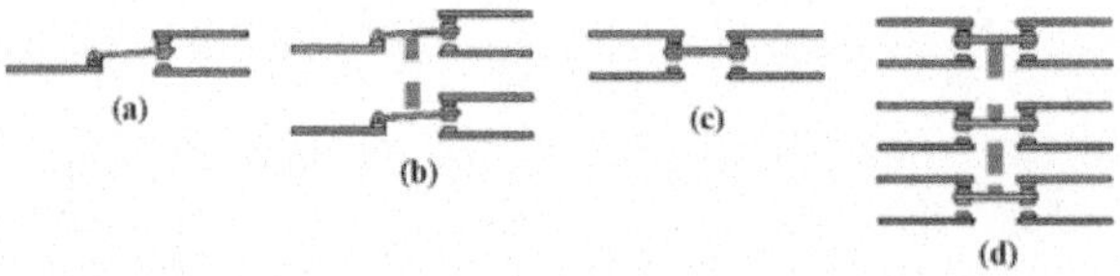

(a) (b) (c) (d)

Figure 7-4 Relay contacts for Question 14.

15. The load-carrying capacity of contacts is normally given as a current value for a _______________.

    a. Resistive load

    b. Inductive load

    c. Capacitive load

    d. Shorted load

    Answer _______

16. A relay contact rated 12 A at 12 V DC can switch _______________ A at 48 V DC.

17. The term interposing relay refers to a type of relay that enables the energy in a _______________ circuit to be switched by a low-power control signal.

PART 2 Quiz: Solid-State Relays

1. In the symbol of the solid-state relay (SSR) shown in Figure 7-5, the #1 set of leads is comparable to the _______________ leads of an electromechanical relay and the #2 set to the _______________ leads.

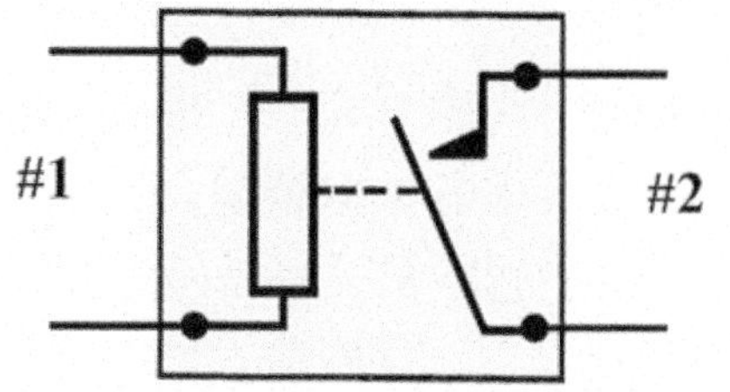

Figure 7-5 Symbol for Question 1.

2. List three common main switching semiconductors used in solid-state relays.
   a. _______________________________
   b. _______________________________
   c. _______________________________
3. A common method used to provide isolation is to have the input section illuminate a _______________ that activates a photodetective device.
   a. LED
   b. SCR
   c. Transistor
   d. Triac

   Answer _______
4. Identify the components of the SSR shown in Figure 7-6.
   a. _______________________________
   b. _______________________________
   c. _______________________________
   d. _______________________________

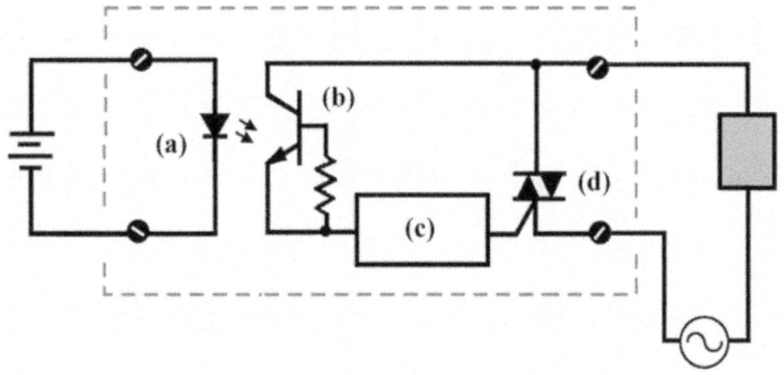

Figure 7-6 Solid-state relay for Question 4.

5. The input section of a SSR acts like the contacts of an electromagnetic relay (EMR). (True/False) _______________
6. The use of solid-state relays is not recommended for loads that must be switched continually and quickly. (True/False) _______________
7. Most SSRs have a variable input voltage range. (True/False) _______________
8. A majority of SSRs are multiple-pole devices. (True/False) _______________
9. Complete the solid-state relay wiring diagram in Figure 7-7 to operate so that when the control contact is closed, three-phase power is delivered to the load.

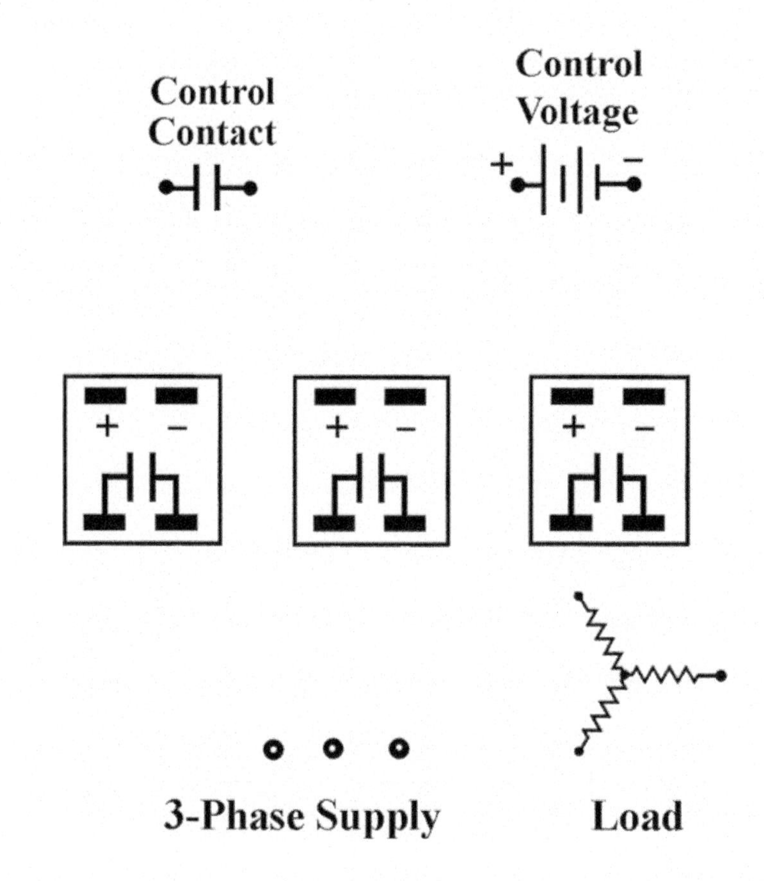

Figure 7-7 Wiring diagram for Question 9.

10. Complete the three-wire motor control diagram for Figure 7-8 utilizing the SCR and solid-state relay. The circuit is to operate so that the starter coil is energized when the start button is momentarily pressed and deenergized whenever either of the two stop buttons is momentarily pressed.

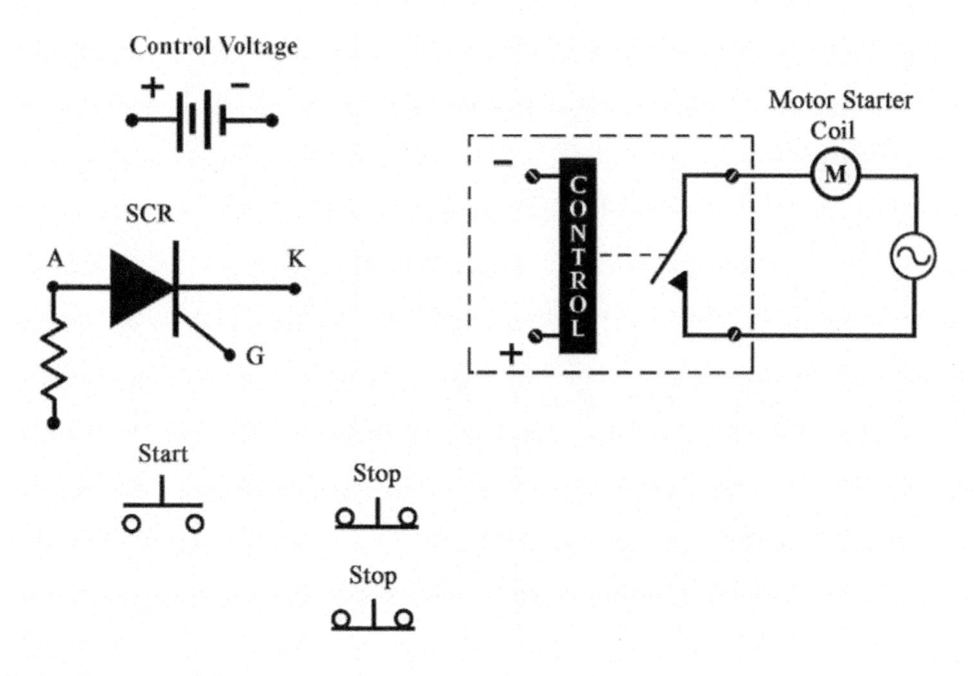

Figure 7-8 Motor control diagram for Question 10.

11. What type of SSR switching is primarily designed to work with resistive loads?
    a. Instant-on switching
    b. Peak switching
    c. Phase-shift switching
    d. Zero switching

    Answer _______

12. _______________ SSR switching initiates the output at the exact same time the input signal is received.

13. Zero switching of resistive loads increases the inrush current.

    (True/False) _______________

14. Compared to electromechanical relays, solid-state relays
    a. Do not exhibit contact bounce
    b. Have a much faster response time
    c. Do not generate as much electromagnetic interference
    d. All of these
    Answer _______
15. Compared to EMR contacts, the SSR switching semiconductor has a much lower on-state resistance. (True/False) _______________
16. Which type of relay has a number of output voltages?
    a. On-delay timer
    b. Off-delay timer
    c. Analog-switching
    d. Digital-switching

PART 3 Quiz: Timing Relays

1. Timing relays are used to _______________ the opening or closing of contacts for circuit control.
2. Contacts of a synchronous clock timer are tripped by
    a. Hand
    b. An electromagnet
    c. Tabs set along a timing wheel
    d. A control signal
    Answer _______
3. Synchronous clock timers are not accurate enough for critical timing processes. (True/False) _______________
4. Timing is managed in a dashpot pneumatic timer by controlling _______________ flow through a small orifice.
5. Pneumatic timers have a relatively narrow timing range. (True/False) _______________
6. Compared with dashpot timers, solid-state timers
    a. Provide much more accurate timing
    b. Have a narrower timing range
    c. Are initiated by an electromagnetic coil
    d. All of these
    Answer _______

7. Complete the solid-state timer wiring diagram for Figure 7-9. The circuit is to operate so that load is energized at 120 V AC following the closure of the float switch and preset time delay period.

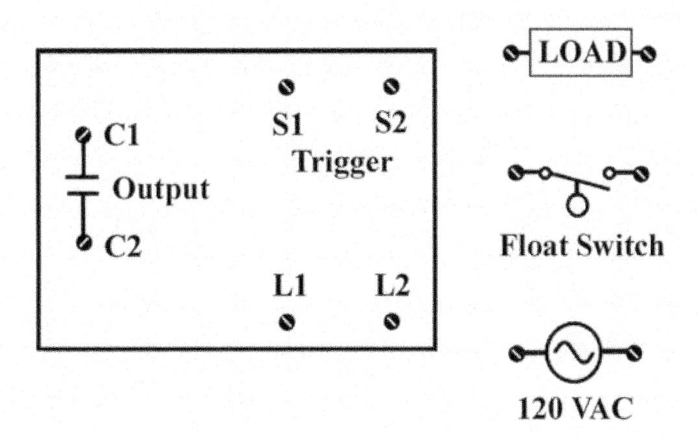

Figure 7-9 Wiring diagram for Question 7.

8. Draw the symbol for the timer contacts listed below.

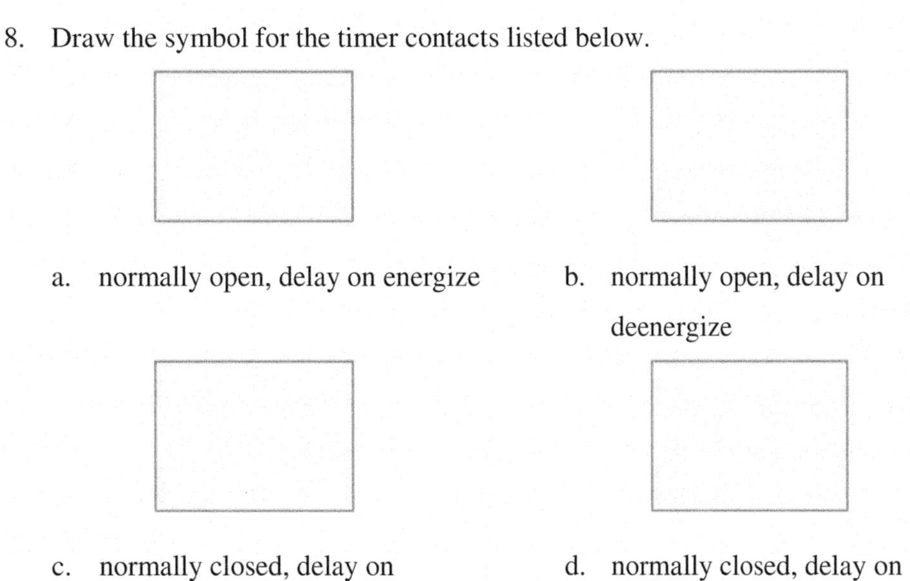

a. normally open, delay on energize

b. normally open, delay on deenergize

c. normally closed, delay on deenergize

d. normally closed, delay on energize

9. Examine the timing diagrams of Figure 7-10 and identify the timer type as DOE, DODE, one-shot, or recycle.

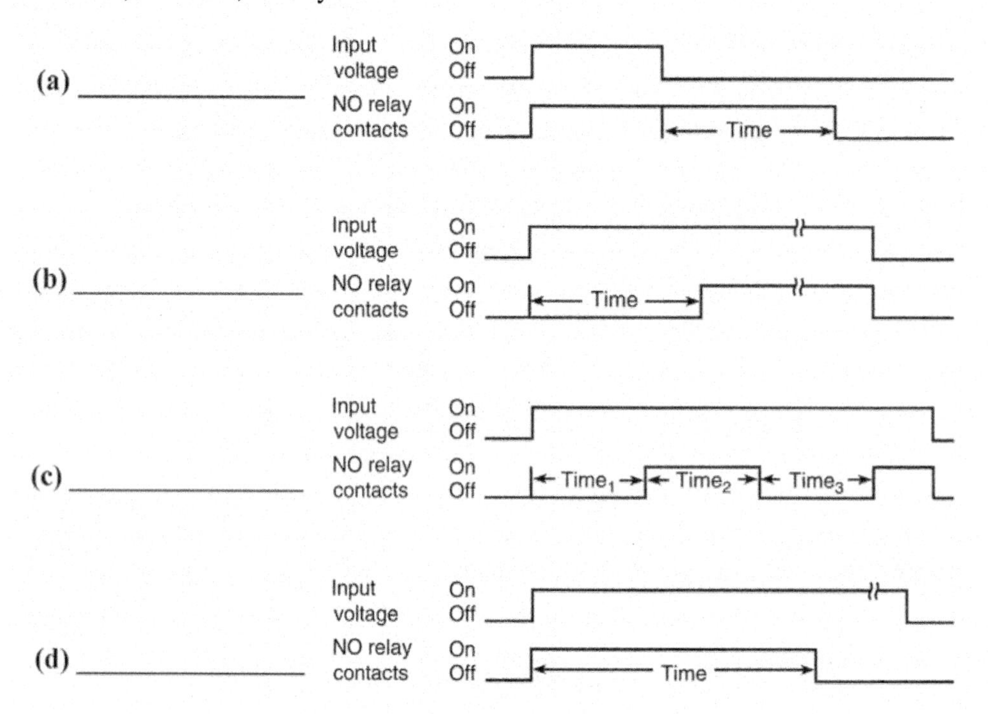

Figure 7-10 Timing diagrams for Question 9.

10. Answer the following with reference to the timer circuit of Figure 7-11.

a. What is the state (open or closed) of contact TR1-1 prior to the switch being closed?

Answer ________

b. What is the state of contact TR1-2 immediately after the switch is closed?

Answer ________

c. What is the state of contact TR1-1 after 5 seconds of the switch being closed?

Answer ________

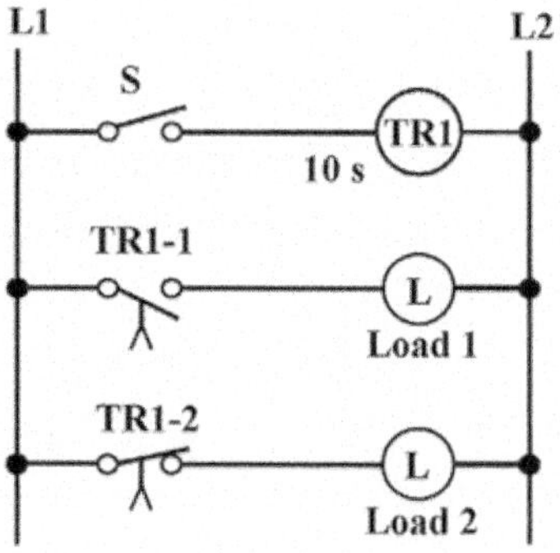

Figure 7-11 Circuit for Question 10.

d. What is the state of contact TR1-2 after 10 seconds of the switch being closed?

Answer ________

e. What is the state of contact TR1-1 as soon as the switch is opened?

Answer ________

11. Answer the following with reference to the timer circuit of Figure 7-12.

    a. What is the state (open or closed) of contact TR1-2 prior to the switch being closed?

    Answer _______

    b. What is the state of contact TR1-1 immediately after the switch is closed?

    Answer _______

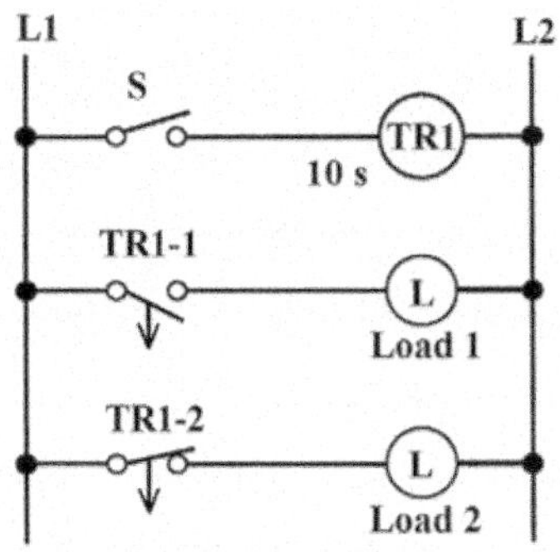

Figure 7-12 Circuit for Question 11.

    c. What is the state of contact TR1-2 immediately after the switch is opened?

    Answer _______

    d. What is the state of contact TR1-1 after 10 seconds of the switch being opened?

    Answer _______

    e. What is the state of contact TR1-2 after 10 seconds of the switch being opened?

    Answer _______

12. Answer the following with reference to the plug-in timer base wiring diagram of Figure 7-13.

    a. What are the pin numbers for the normally open contact?

    Answer _______

    b. What are the pin numbers for the normally closed contact?

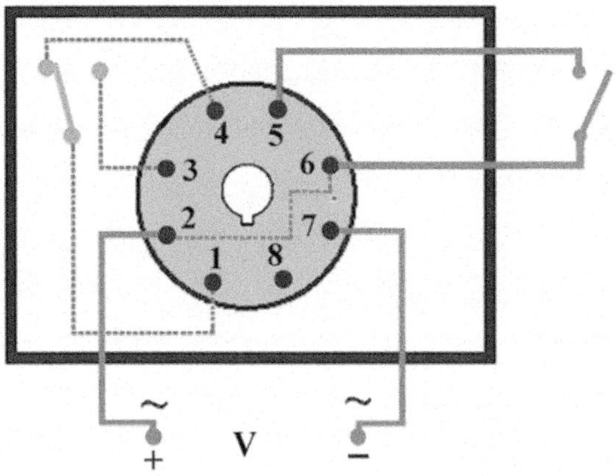

    Answer _______

Figure 7-13 Circuit for Question 12.

    c. Identify the pin numbers where the constant voltage source is to be connected.

    Answer _______

    d. Identify the pin numbers where an external trigger switch is to be connected.

    Answer _______

13. The one-shot timer shown in Figure 7-14 is set for a time-delay period of 5 seconds. For how long would the load be energized for each of the following scenarios?

    a. Initiate button is held closed for 1 second.

    Answer _______

    b. Initiate button is held closed for 5 seconds.

    Answer _______

    c. Initiate button is held closed for 10 seconds.

    Answer _______

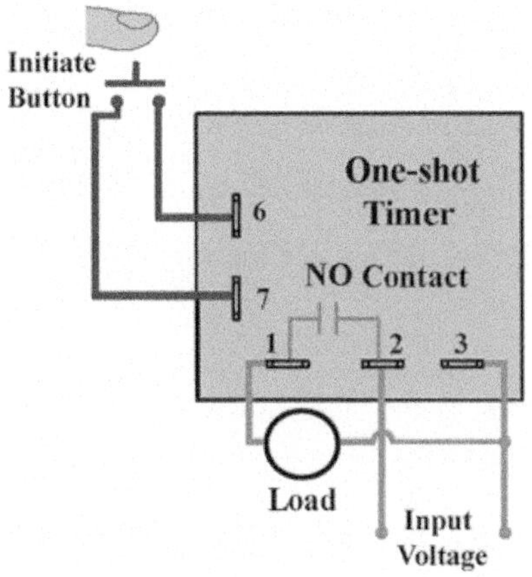

Figure 7-14 Timer for Question 13.

    d. Initiate button is held closed for 2 seconds but power is lost after 1 second.

    Answer _______

14. Answer the following with reference to the recycle timer shown in Figure 7-15.

    a. What type of recycle timer (symmetrical or asymmetrical) is this?

    Answer ______________________________

    b. How is the on period timing adjustment accomplished?

    Answer ______________________________

    c. How is the off period timing adjustment accomplished?

    Answer ______________________________

    d. How is the on/off recycling of the contacts stopped?

    Answer ______________________________

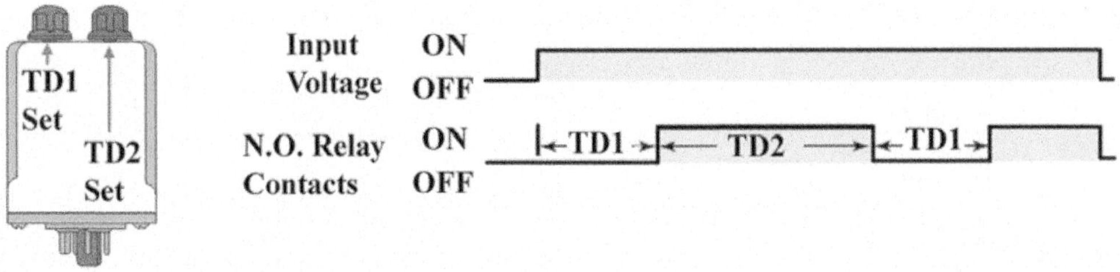

Figure 7-15 Timer for Question 14.

15. The term ______________ timer is reserved for a group of solid-state timers that perform more than one timing function.

16. A programmable logic controller can be programmed to implement conventional timer functions. (True/False) _______________

17. Complete the on-delay ladder logic program of Figure 7-16 to operate so that output Q1 is energized any time both input switches I1 and I3 are closed for a sustained period of 10 seconds or more.

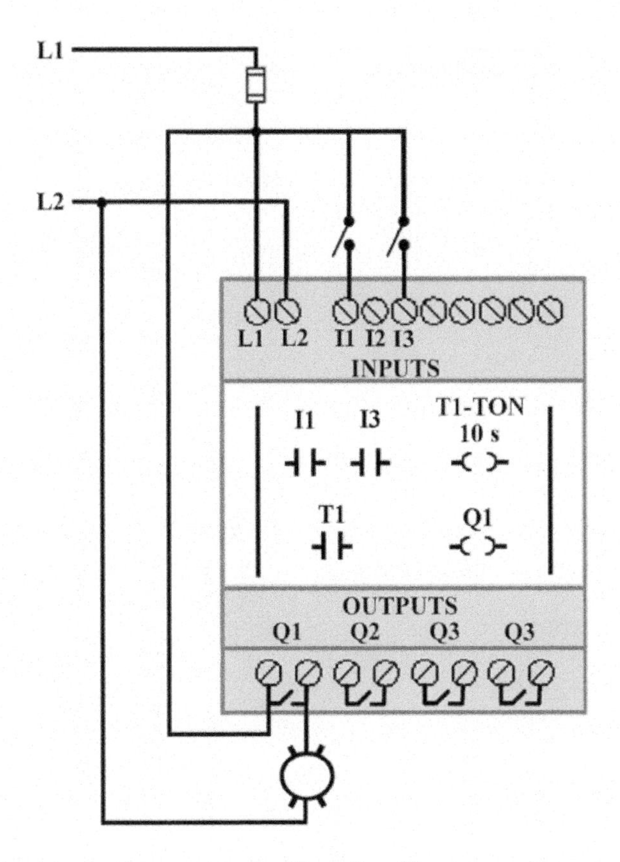

Figure 7-16 Ladder logic program for Question 17.

18. A recycle timer can be used in a flasher circuit used to operate a waring pilot light on and off. (True/False) _______________

19. A one-shot timer is used to have a momentary input trigger a device or operation for a preset period of time. (True/False) _______________

20. Which type of timer relay contacts are controlled directly by the timer coil, as in a general-purpose control relay?

    a. Normally open
    b. Normally closed
    c. Timed
    d. Instantaneous

PART 4 Quiz: Latching Relays

1. Latching relays use a _______________ latch or permanent _______________ to hold the contacts in their last energized position.
2. In a latching relay that uses two coils, the coils are identified as the _______________ coil and the _______________ coil.
3. Latching relays can provide circuit continuity during power failures. (True/False) _______________
4. In a two-coil latching relay, the latch coil requires a _______________ flow of current to set the latch and hold the relay in the latched position.
   a. Direct
   b. Alternating
   c. Continuous
   d. Pulsed

   Answer _______
5. On electrical diagrams, the contacts of a latching relay are normally shown
   a. In latched position
   b. In unlatched position
   c. Energized
   d. Deenergized

   Answer _______
6. Answer the following with reference to the latching relay circuit of Figure 7-17.
   a. How is the pilot light turned on?

   Answer ____

   b. How is the pilot light turned off?

   Answer ____

c. Assume the unlatch coil is faulted open. How would this affect the operation of the circuit?

Answer ____

____

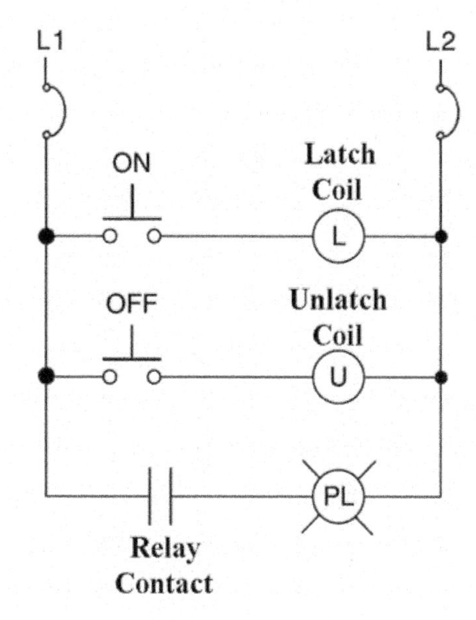

Figure 7-17 Circuit for Question 6.

7. Single-coil magnetic latching relays are designed to be polarity sensitive. (True/False) ______________

8. Alternating relays are used to equalize the ______________ ______________ of two loads.

9. Answer the following with reference to the alternating relay circuit of Figure 7-18.

   a. What configuration of relay contacts is used?

   Answer ______________________________

   b. What initiates the switch in output between Load A (LA) and Load B (LB)?

   Answer ______________________________

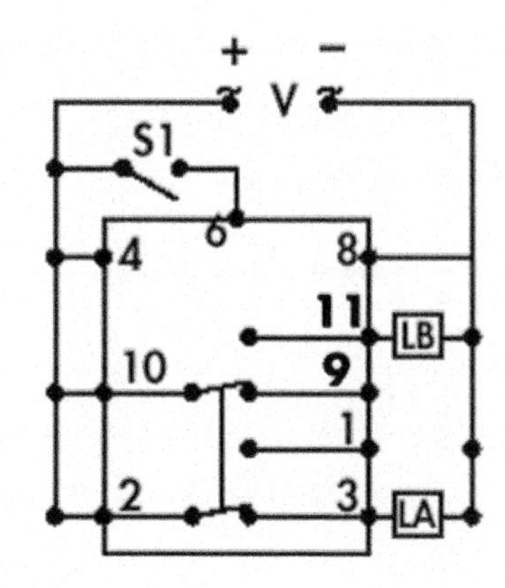

Figure 7-18 Relay circuit for Question 9.

c. Remote pilot lights are to be connected to the unused contacts of the relay to indicate which load is active. To which two pins would the pilot lights for load A (LA) and load B (LB) be connected?

Answer ____

10. ______________ - ______________ alternating relays have the ability to alternate the loads during normal work loads and then operate both when the load is high.

PART 5 Quiz: Relay Control Logic

1. Identify the two states of the digital signal shown in Figure 7-19.

   a. ______________ b. ______________

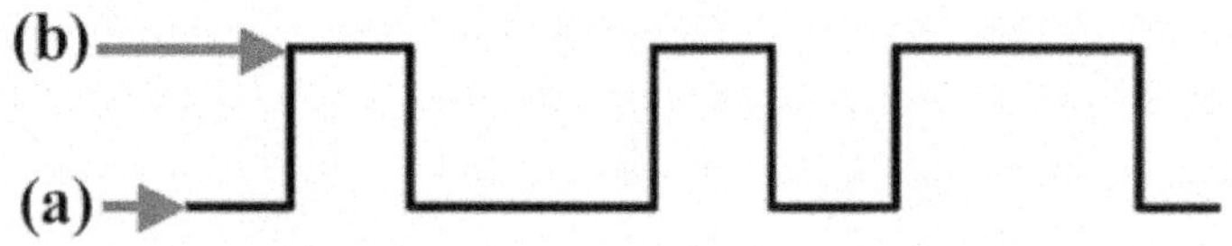

Figure 7-19 Digital signal for Question 1.

2. Relay control logic and ladder diagrams are the primary programming language for ____.

3. The two separate and distinct parts of a motor control circuit are the primary and secondary sections. (True/False) ______________

4. The input section of a control circuit provides the

   a. Signals

   b. Source

   c. Actions

   d. Load

   Answer _______

5. Answer the following with reference to the control circuit of Figure 7-20.

   a. List all the component parts that are considered to be inputs.

   Answer

   ____

   ____

   ____

   ____

   b. List all the component parts that are considered to be outputs.

   Answer

   ____

   ____

   ____

   ____

   ____

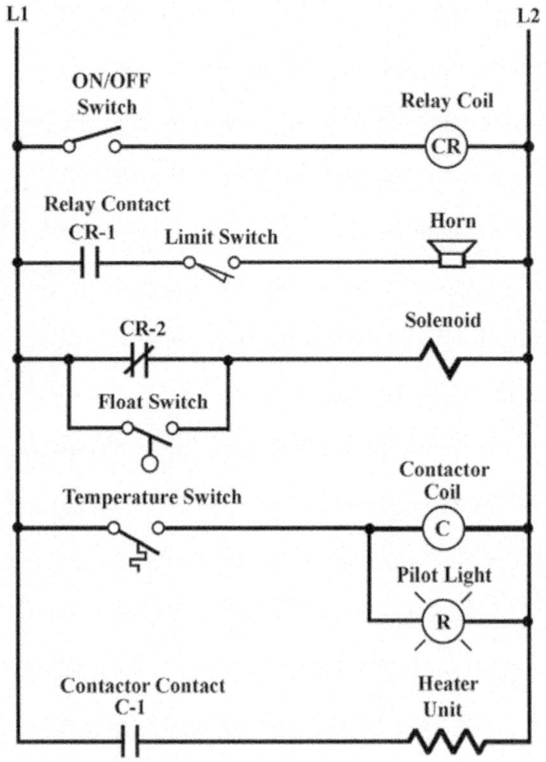

Figure 7-20 Control circuit for Question 5.

   c. Which output would be energized with power applied and the circuit in the state shown?

   Answer _______

   d. Assume that the contactor coil becomes faulted open. In what way would the normal operation of the circuit be affected when the temperature switch closes?

   Answer ____

   ____

6. All input devices can be classified as being loads. (True/False) _______________

7. Logic functions describe how inputs interact with each other to control the outputs. (True/False) _______________

8. Logic is the ability to make decisions when one or more different factors must be taken into consideration. (True/False) _______________

9. Identify the control logic function associated with each of the contact configurations shown in Figure 7-21.

a. _______________ b. _______________ c. _______________

d. _______________ e. _______________ f. _______________

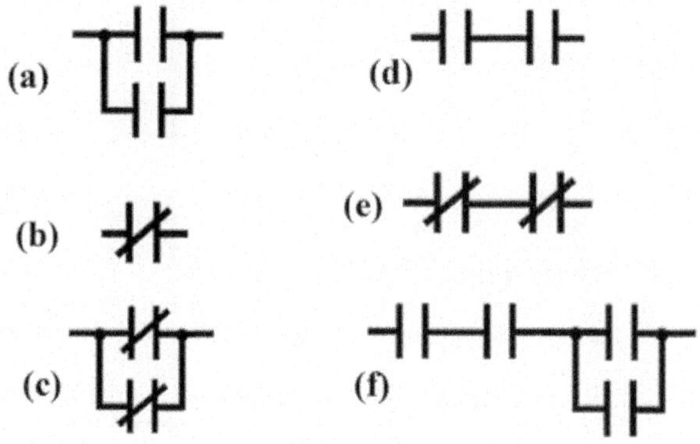

(a)

(b)

(c)

(d)

(e)

(f)

Figure 7-21

Configurations for Question 9.

10. State the type of logic function described in each of the following control circuit scenarios.

    a. Two or more inputs are connected in series and they all must be closed to energize the load.

    Answer _______

    b. Two or more inputs are connected in parallel and any one of the inputs can close to energize the load.

    Answer _______

11. Draw an OR logic circuit in which a normally open push button and temperature switch control a resistive heating element.

12. Draw a NOT logic circuit in which a solenoid and pilot light are controlled by a single push button. Both loads are to be energized when the button is not pressed and deenergized when the button is pressed.

13. Draw a combination NOR/OR logic circuit that uses a three-wire control to energize a magnetic motor starter. The motor can be started and stopped from three locations. Include overload protection for the motor.

hands on PRACTICAL ASSIGNMENTS

1. The purpose of this assignment is to wire different types of relays in simple configurations that demonstrate their basic operating principle.
   a. Complete a circuit diagram for an electromechanical relay configured to operate a lamp load. Wire the circuit in a neat and professional manner.
   b. Complete a circuit diagram for a solid-state relay configured to operate a lamp load. Wire the circuit in a neat and professional manner.
   c. Complete a circuit diagram for a timing relay configured to operate a lamp load. Wire the circuit in a neat and professional manner.
   d. Complete a circuit diagram for an off-delay latching relay configured to operate a lamp load. Wire the circuit in a neat and professional manner.
   e. Complete a circuit diagram for an alternating relay configured to operate a lamp load. Wire the circuit in a neat and professional manner.
2. The purpose of this assignment is to wire, program, and observe the operation of a programmable logic controller (PLC) on-delay timer.
   a. Construct a PLC wiring diagram and program using whatever type PLC is available to you and configured to operate a lamp load.
   b. Complete the hard wiring of the circuit.
   c. Enter and run the program.

8 Motor Control Circuits

PART 1 Quiz: NEC Motor Installation Requirements

Place the answers in the space provided.

1. Article _______________ of the National Electrical Code covers installation of motor circuits.
2. Using the list of terms provided below, identify by letter each requirement for the motor installation shown in Figure 8-1.
   a. Overload protection _______________
   b. Short-circuit protection _______________
   c. Control circuit protection _______________
   d. Motor controller _______________
   e. Disconnecting means _______________
   f. Sizing branch circuit conductors _______________
   g. Providing a control circuit _______________

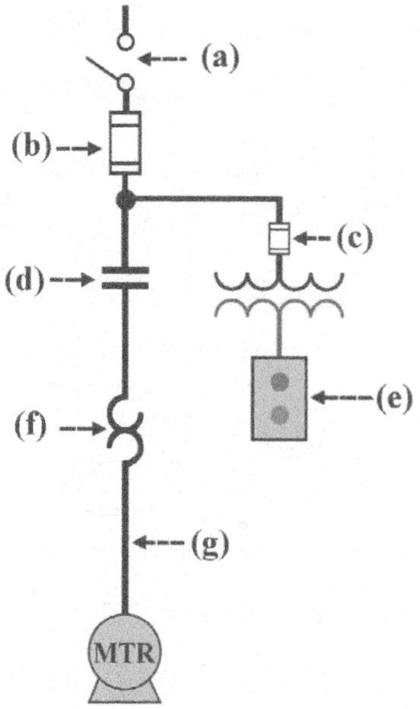

(a)
(b)
(c)
(d)
(e)
(f)
(g)
MTR

Figure 8-1 Motor installation for Question 2.

3. Motor branch circuit conductors must have an ampacity of not less than _______________ of the motor's full-load current.
   a. 50 percent
   b. 100 percent
   c. 125 percent
   d. 200 percent

   Answer _______

4. The full-load current rating shown on the motor's nameplate is not permitted to be used to determine the ampacity of the conductors. (True/False) _______________

5. The full-load current rating shown on the motor's nameplate is permitted to be used to determine the overload protection for the motor. (True/False) ______________

6. According to the NEC, determine the minimum branch-circuit ampacity required for a 7½-hp, 230-V, three-phase motor.

   Answer _______

7. A 20-hp motor and a 10-hp motor, both 460 V, three phase, share the same feeder. Determine the minimum branch circuit ampacity required to size the feeder conductors.

   Answer _________

8. For the motor branch circuit shown in Figure 8-2, state what undesirable operating condition each of the protective devices guards against.

   a. ____

   b. ____

   c. ____

   d. ____

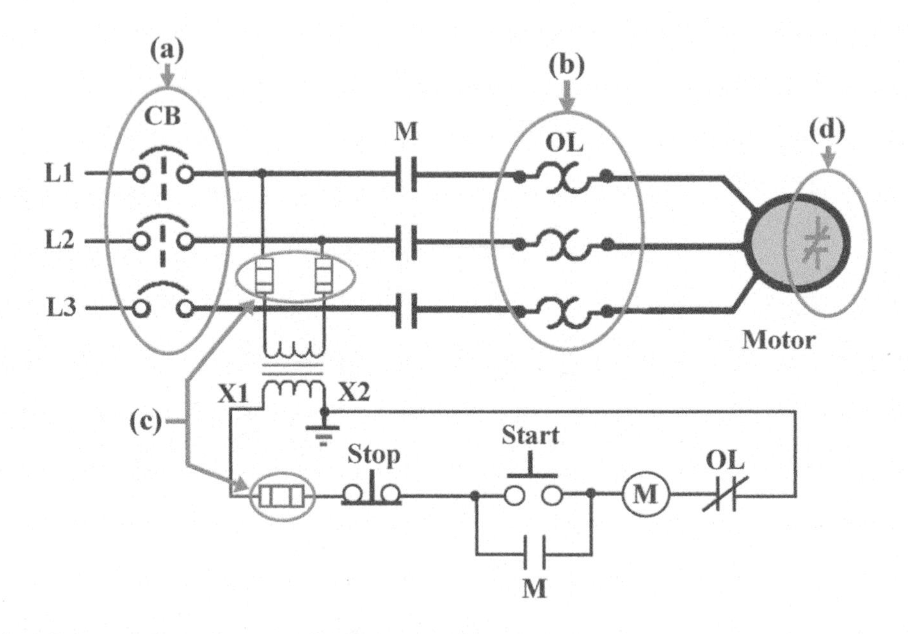

Figure 8-2 Branch circuit for Question 8.

9. The starting current of a motor is approximately the same as its rated full-load current. (True/False) ______________

10. With an inverse time circuit breaker, the higher the overcurrent, the longer the time required for the breaker to trip. (True/False) ______________

11. According to the NEC, what size of inverse time circuit breaker is permitted to be used for short-circuit and ground fault protection for a 5-hp, 230-V, three-phase, squirrel-cage motor?

    Answer _______

12. A motor overload condition is normally caused by

    a. A ground fault
    b. A short circuit
    c. An open equipment grounding conductor
    d. Excessive motor load

13. Time-delay fuses provide both overload and short-circuit protection. (True/False) _______________

14. For any given horsepower motor, the nameplate full-load ampere rating is the same as that given in NEC tables. (True/False) _______________

15. In the majority of applications, motor overload protection is provided by overcurrent _______________ in the motor controller.

16. Motor control circuit conductors are not permitted to be tapped from the motor branch circuit conductors. (True/False) _______________

17. Motor control circuits do not carry the main power current to the motor. (True/False) _______________

18. A motor _______________ is used to start and stop an electric motor by closing and opening the main power current to the motor.

19. Which of the following is an example of a motor controller?

    a. Magnetic starter
    b. Manual snap switch
    c. Adjustable-speed drive
    d. All of these

    Answer _______

20. The purpose of a motor _______________ means is to open the supply conductors to the motor, allowing personnel to work safely on the installation.

21. The disconnecting means for the motor controller and the motor must open all _______________ supply conductors simultaneously.

22. In general, each motor requires a disconnecting means that is located within sight of the motor. "In sight of" is defined as being visible and not more than _______________ feet in distance from one another.

23. An isolation switch should not be operated under load. (True/False) _______________

24. Motor control circuits can be the same voltage as the motor up to 600 V or can be reduced by means of a control _______________.

25. Where one side of the motor control circuit is grounded, the design of the control circuit must prevent the motor from being started because of a _______________ fault in the control circuit wiring.

PART 2 Quiz: Motor Starting

1. All motors tend to draw much more current during the starting period than when rotating at operating speed. (True/False) _______________

2. The current that flows before a motor begins to turn is called its
   a. Locked-rotor current
   b. Full-load current
   c. No-load current
   d. Rated current

   Answer _______

3. An across-the-line starter is designed to apply full line voltage to the motor upon starting. (True/False) _______________

4. Manual starters cannot be operated on/off by remotely located control devices. (True/False) _______________

5. The starter shown in Figure 8-3 would be classified as which of the following types?
   a. Single-pole manual
   b. Double-pole magnetic
   c. Three-pole magnetic
   d. Three-pole manual

   Answer _______

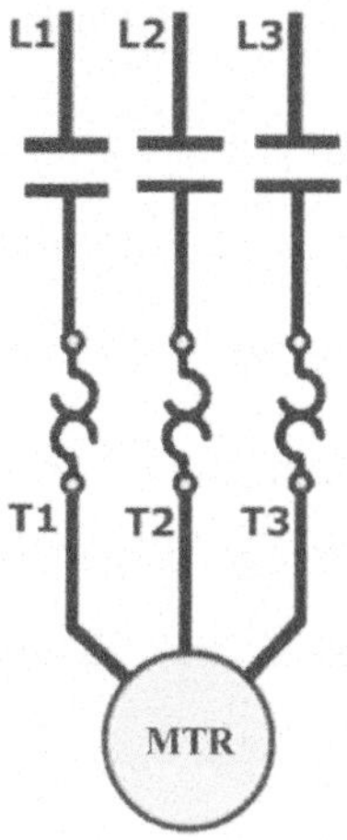

Figure 8-3 Starter for Question 5.

6. On a combination starter, the door cannot be opened while the disconnecting means is closed. (True/False) _______________

7. Complete a wiring diagram for the magnetic across-the-line starter shown in Figure 8-4, according to the two start/stop station line diagrams shown in the text. Do not make any wire splices, and run all wire connections from one terminal screw to the other.

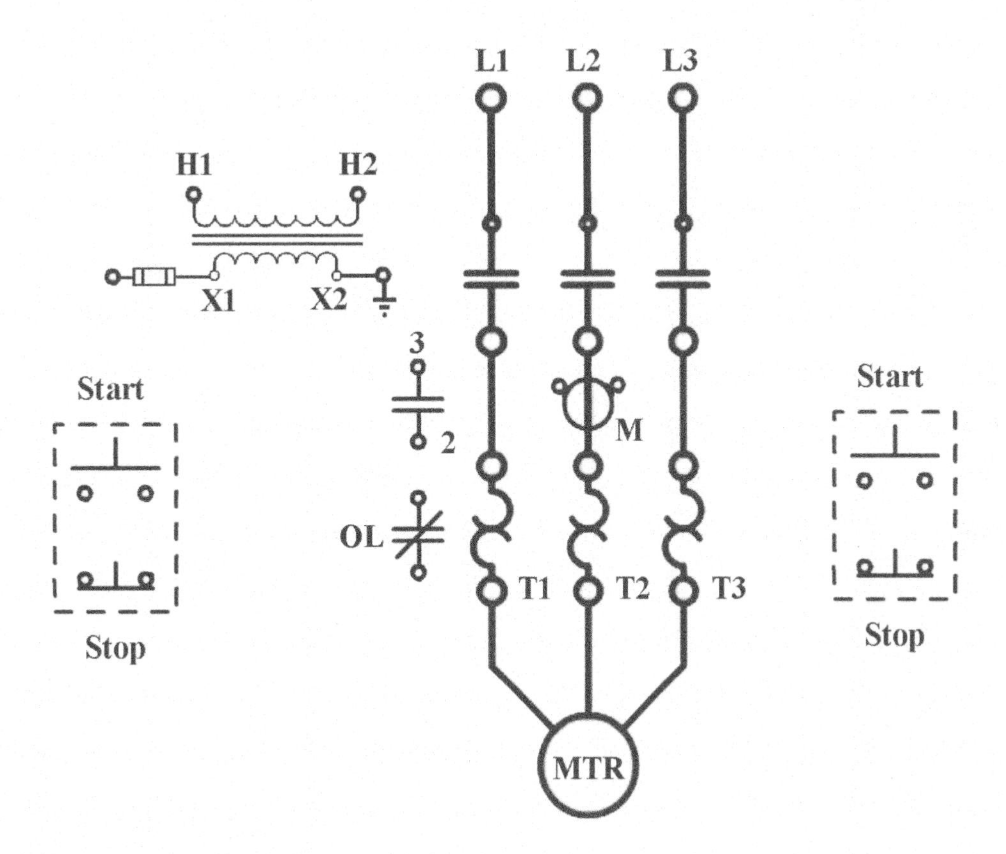

Figure 8-4 Starter for Question 7.

8. A reduced voltage starter
   a. Limits line voltage disturbances
   b. Reduces torque on starting
   c. Increases torque on starting
   d. Both a and b

   Answer _______

9. Answer the following with reference to the primary resistance starter shown in Figure 8-5.

   a. Pressing the start button energizes coils ______________ and ______________.

   b. The motor is initially started through a ______________ in each of the incoming lines.

   c. After a preset time delay period, the ______________ contact closes to energize coil ______________, the contacts of which close to apply full voltage to the motor.

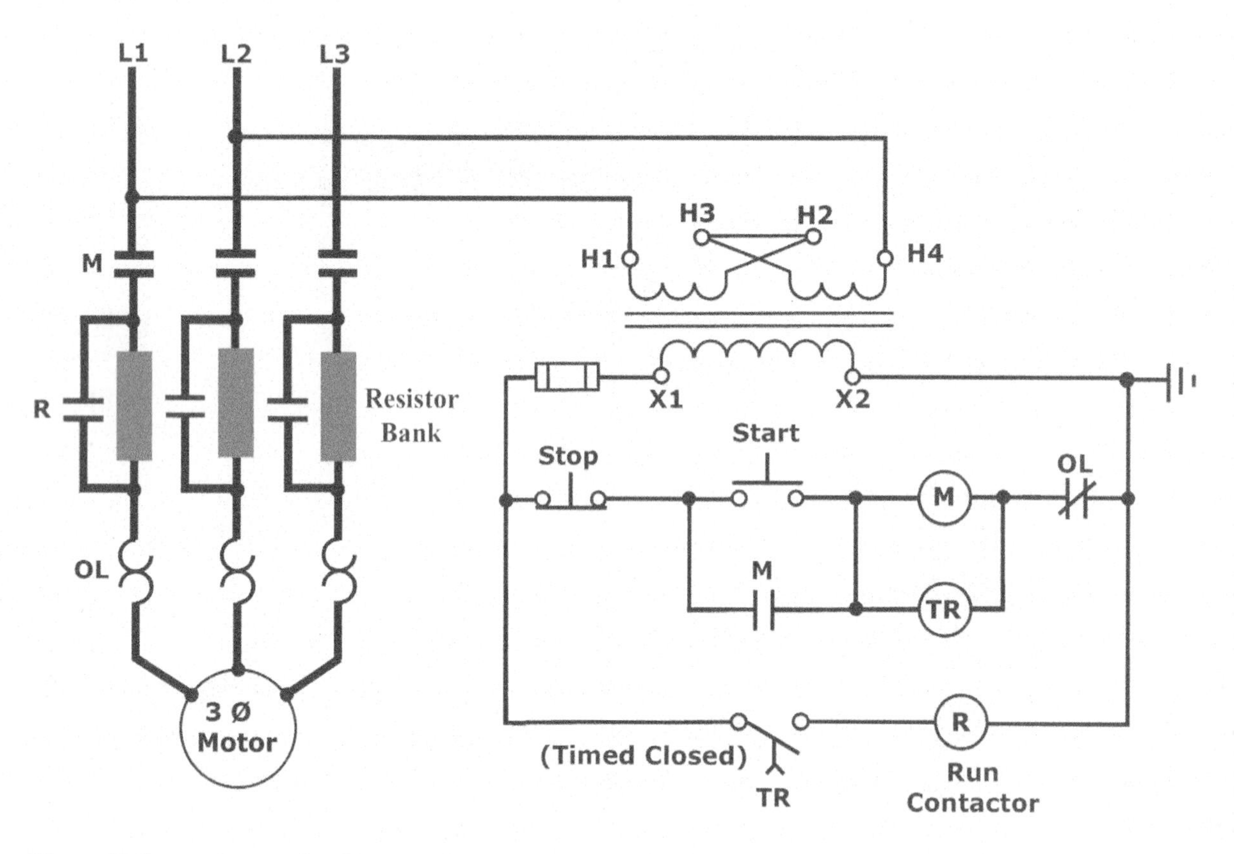

Figure 8-5 Starter for Question 9.

10. Answer the following with reference to the autotransformer starter shown in Figure 8-6.

    a. The instantaneous contacts of timer coil TR are ______________ and ______________, while the timed contacts are ______________ and ______________.

b. The net result of initially pressing the start button is to energize coils

_______________, _______________, and _______________.

c. After the timed contacts change state, coils _______________ and

_______________ deenergize and coil _______________ energizes,

connecting the motor directly to the _______________, ___________ voltage.

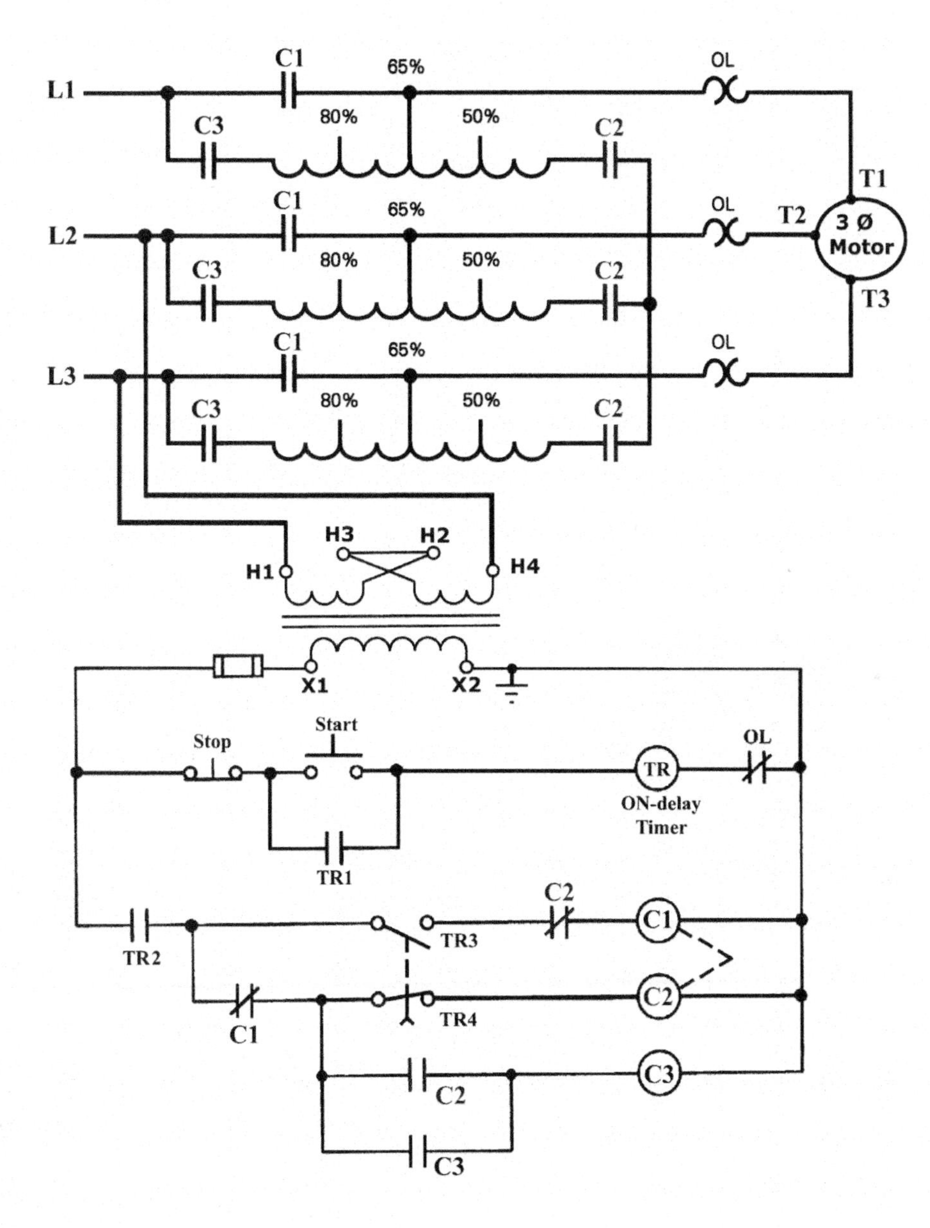

Figure 8-6 Starter for Question 10.

11. Wye-delta starting involves connecting the motor windings in a _______________ configuration during the starting period and then in _______________ after the motor has begun to accelerate.
12. Wye-delta starters can be used with motors where three or six leads of the stator winding are available. (True/False) _______________
13. Open transition wye-delta starters
    a. Apply voltage to the motor during transition
    b. Do not apply voltage to the motor during transition
    c. Do not create voltage surges during transition
    d. Both b and c

    Answer _______
14. Part-winding reduced-voltage starters are used on three-phase motors wound for _______________-voltage operation.
15. Part-winding starters are designed to start the motor by energizing both sets of windings on starting and then disconnecting one set when the motor nears operating speed. (True/False) _______________
16. Which of the following starters increases the voltage gradually as the motor starts?
    a. Autotransformer starter
    b. Solid-state starter
    c. Wye-delta starter
    d. All of these

    Answer _______
17. Soft-start motor starters limit motor starting current and torque by ramping the voltage applied to the motor during the starting. (True/False) _______________
18. Which of the following soft-start acceleration modes of operation regulates the maximum starting current to some preset value?
    a. Dual ramp start
    b. Current limit start
    c. Linear speed acceleration
    d. Full-voltage start

    Answer _______

19. Answer the following with reference to the DC motor starter shown in Figure 8-7.

   a. This starter would be classified as a reduced-voltage type.

      (True/False) _______________

   b. The use of the three power contacts connected in series is designed to reduce contact _______________.

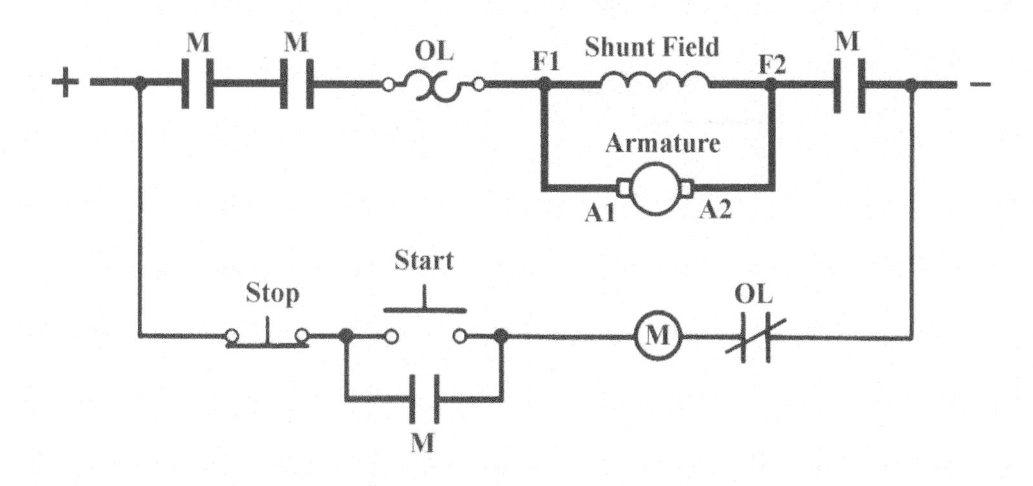

Figure 8-7 Starter for Question 19.

20. Answer the following with reference to the DC motor starter shown in Figure 8-8.

   a. This starter would be classified as an across-the-line type.

      (True/False) _______________

   b. The voltage applied to the L1, L2, and L3 inputs is three-phase AC.

      (True/False) _______________

   c. The function of the SCRs is to _______________ the AC input voltage and provide a _______________ DC voltage to the armature.

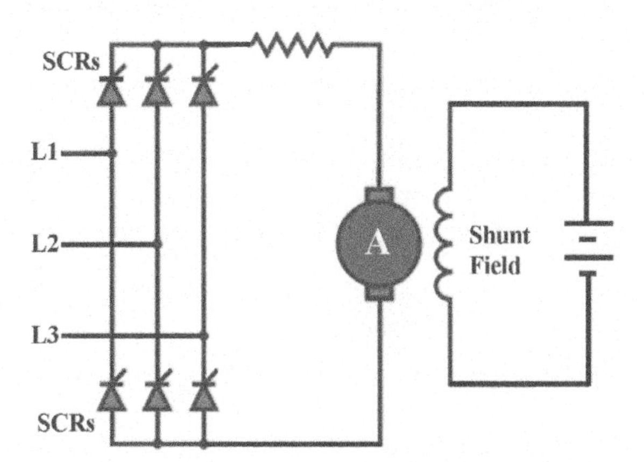

Figure 8-8 Starter for Question 20.

21. A forward-reverse motor starter must be prevented from form reversing while running in the forward rotation. What is done?
    a. Install electrical limit switches.
    b. Install electrical interlocking.
    c. Install a NC stop button.
    d. Install a NO pushbutton.
22. With reference to the circuit shown, when will the horn sound?
    a. Horn is ON for 10 seconds before the motor starts.
    b. Horn is ON for 10 seconds after the motor starts.
    c. Horn is ON for 10 seconds before the motor stops.
    d. Horn is ON for 10 seconds after the motor stops.

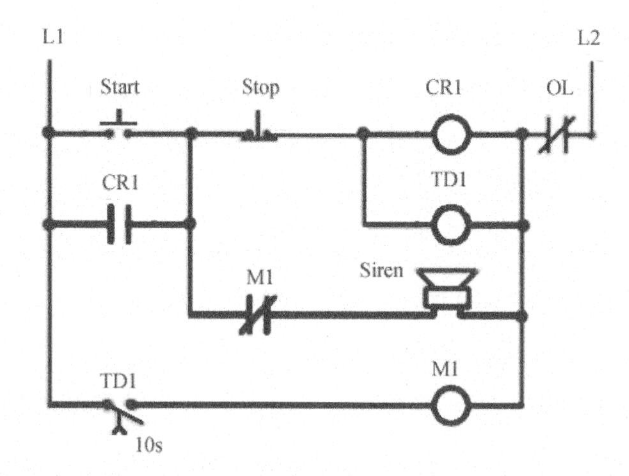

23. What type of contact is an emergency stop button?
    a. Momentary, NC
    b. Momentary, NO
    c. Latching, NO
    d. Latching, NC

PART 3 Quiz: Motor Reversing and Jogging

1. Interchanging any two line leads to a three-phase induction motor will reverse its direction of rotation. (True/False) _______________
2. A reversing starter requires _______________ set(s) of overload relays.
3. A reversing starter requires two contactors. (True/False) _______________

4. In a three-phase reversing starter, the forward and reverse contactors are mechanically interlocked in order to
   a. Allow the motor to accelerate to full speed
   b. Prevent chattering of the starter coil
   c. Prevent a short circuit from being created across the lines
   d. Both b and c

   Answer _______

5. Pushbutton interlocking of reversing motor starters utilizes
   a. A mechanical link between the forward and reverse push buttons
   b. Momentary and maintained switch contacts
   c. Break-make switch contacts
   d. All of these

   Answer _______

6. With pushbutton interlocking, the normally closed contact on the reverse push button acts like another _______________ push button in the forward circuit.

7. Answer the following with reference to the reversing starter control circuit shown in Figure 8-9. This circuit incorporates limit switches to limit travel in the forward and reverse directions.

Figure 8-9 Starter for Question 7.

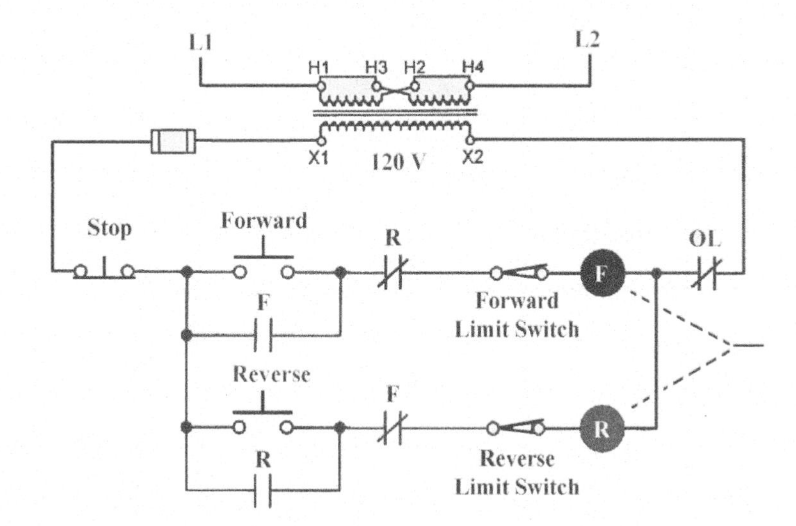

   a. Pressing the forward push button energizes coil _______________.
   b. Once operating in the forward direction, the motor is stopped by actuation of the _______________ push button or the _______________ limit switch.

c. Both limit switches would be classified as being normally

   _______________.

d. With both limit switches closed and the motor operating in the forward direction, pressing the reverse push button will change the direction of rotation. (True/False) _______________

8. The reversing starter for a single-phase capacitor start motor changes the direction of rotation by interchanging the _______________ winding leads, while those of the run winding remain the same.
9. As a result of the operation of the _______________ switch, the single-phase capacitor-start motor must be allowed to slow down before you attempt to reverse the direction of rotation.
10. The reversal of a DC motor is accomplished by
    a. Reversing the two line leads to the motor
    b. Reversing the direction of the armature current only
    c. Reversing the direction of the field current only
    d. Either b or c

    Answer _______
11. The speeding up of a motor when unloaded is known as jogging. (True/False) _______________
12. When utilizing jogging control
    a. Care must be taken to prevent dynamic braking
    b. The maintaining contact must not be connected
    c. The thermal-overload element must be removed
    d. All of these

    Answer _______

13. Answer the following with reference to the jog circuit shown in Figure 8-10.

   a. This jog circuit would be classified as a control relay type. (True/False) _______________

   b. When the switch is placed in the _______________ position, the maintaining circuit is not broken.

   c. The dual-voltage primary of the control transformer is connected for the higher of the two input voltage ratings. (True/False) _______________

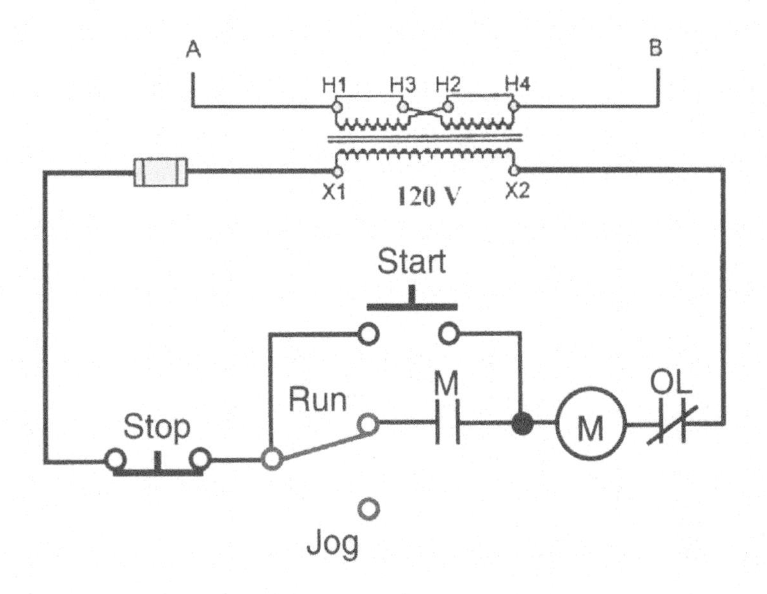

Figure 8-10 Jog circuit for Question 13.

14. What type of contact is an emergency stop button?

   a. Momentary, NC

   b. Momentary, NO

   c. Latching, NO

   d. Latching, NC

   Answer _______

PART 4 Quiz: Motor Stopping

1. The most common method of stopping a motor is to remove the _______________ voltage and allow the motor to coast to a stop.
2. Which of the following braking systems applies to stopping a motor by reversing it while it is still running in the forward direction?
   a. Dynamic braking
   b. Regenerative braking
   c. Plugging
   d. Jogging

   Answer _______
3. Which of the following braking systems can hold a motor stationary once it has stopped?
   a. Dynamic braking
   b. Eddy-current brake
   c. Friction brake
   d. Both a and b

   Answer _______
4. A _______________ starter is required when a motor is to be brought to a stop by means of plugging.
5. Plugging produces more heat in both the motor and starter than allowing a motor to coast to a stop. (True/False) _______________
6. A zero-speed or plugging switch prevents the motor from _______________ after it has come to a stop.
7. Answer the following with reference to the antiplugging protection circuit shown in Figure 8-11.
   a. In this application the motor can be reversed but not plugged. (True/False) _______________
   b. With the motor operating in the forward direction, the F zero-speed switch contacts will be _______________ and the R zero-speed contacts will be _______________.

c. With the motor operating in the reverse direction, the F zero-speed switch contacts will be _______________ and the R zero-speed contacts will be _______________.

d. With the motor at zero speed the F zero-speed switch contacts will be _______________ and the R zero-speed contacts will be _______________.

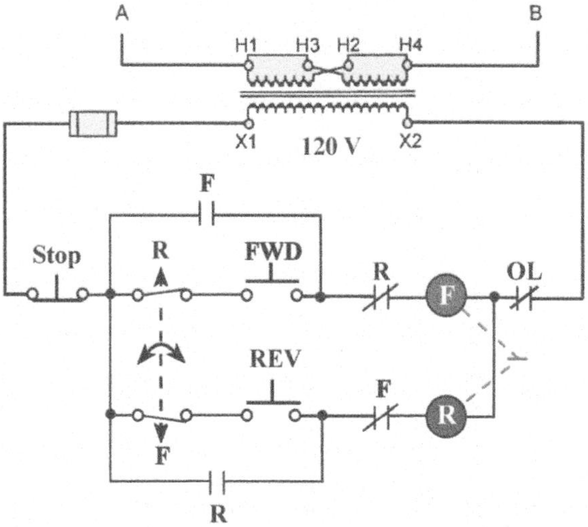

Figure 8-11 Antiplugging circuit for Question 7.

8. Dynamic braking is achieved by reconnecting a rotating motor to act as a _______________ immediately after it is turned off.

9. DC injection braking is a method of braking in which
   a. DC is applied to the stator windings of an AC motor
   b. AC is applied to the stator windings of an AC motor
   c. DC is applied to the rotor windings of an AC motor
   d. AC is applied to the rotor windings of an AC motor

   Answer _______

10. Electromechanical friction brakes rely on fiction between a _______________ and _______________ brake arrangement.

11. Electromechanical friction braking is applied by energizing and deenergizing the brake _______________.

PART 5 Quiz: Motor Speed

1. The speed of an induction motor depends on the
   a. Number of poles built into the motor
   b. Frequency of the power supplied to it
   c. Type of braking system employed
   d. Both a and b

   Answer _______

2. A single-speed motor has one rated speed at which it runs when supplied with the nameplate voltage and frequency. (True/False) _______________

3. Multispeed motors are available in two basic versions: _______________ pole and _______________ winding.

4. Answer the following with reference to the multispeed motor circuit shown in Figure 8-12.

   a. The multispeed motor shown would be classified as a _______________ _______________ type.

   b. This motor requires _______________ standard three-pole starter units.

   c. The starters must be mechanically and electrically _______________ with each other.

   d. Only one set of overload relays is required to ensure protection on each speed range. (True/False) _______________

   e. A six-lead induction motor is required for this application. (True/False) _______________

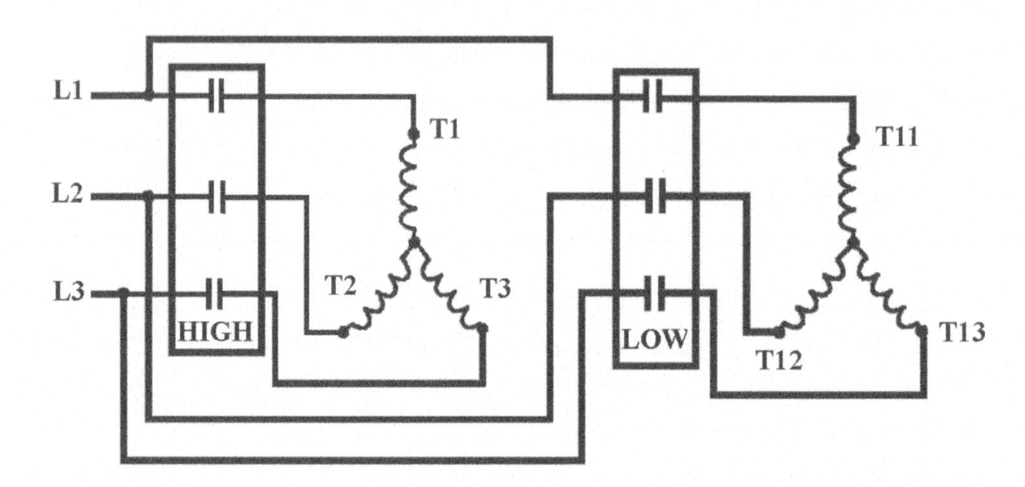

Figure 8-12 Multispeed motor circuit for Question 4.

5. A wound-rotor induction motor has a _______________ like the squirrel-cage motor, but a _______________ with windings brought out via slip rings and brushes.

6. The speed of wound-rotor induction motor is varied by
   a. Connecting variable resistance into the rotor circuit
   b. Connecting variable resistance into the stator circuit
   c. Applying variable voltage to the rotor circuit
   d. Both a and c

   Answer _______
7. The lower the rotor resistance of a wound-rotor motor, the higher the speed of the motor. (True/False) ______________

hands on PRACTICAL ASSIGNMENTS

1. The purpose of this assignment is to wire and operate different types of motor starters. Wire as many of the following motor-starting circuits as possible, using whatever type of motor control station is available to you. Complete a line diagram and a wiring diagram for each circuit and have them checked by the instructor. Wire each circuit in a neat and professional manner and have the installation evaluated by the instructor.
   - Single-phase, fractional-horsepower manual starter control of an AC motor.
   - Three-phase manual control of an AC motor.
   - Three-phase magnetically operated across-the-line starter with a single start-stop/station.
   - Three-phase magnetically operated across-the-line starter with two start/stop stations.
   - Reduced-voltage primary-resistance starter.
   - Reduced-voltage autotransformer starter.
   - Reduced-voltage wye-delta starter.
   - Reduced-voltage part-winding starter.
   - Solid-state soft-start starter.
   - DC across-the-line starter.

2. The purpose of this assignment is to wire and operate different types of motor reversing and jogging circuits. Wire as many of the following motor-reversing and jogging circuits as possible, using whatever type of motor control station is available to you. Complete a line diagram and a wiring diagram for each circuit and have them checked by the instructor. Wire each circuit in a neat and professional manner and have the installation evaluated by the instructor.
   - Across-the-line reversing starter with mechanical interlocking.
   - Across-the-line reversing starter with mechanical and electrical auxiliary contact interlocking.
   - Across-the-line reversing starter with mechanical and electrical pushbutton interlocking.
   - Three-phase across-the-line starter with jog, start, and stop control
3. The purpose of this assignment is to wire and operate different types of motor-stopping circuits. Wire as many of the following motor-stopping circuits as possible, using whatever type of motor control station is available to you. Complete a line diagram and a wiring diagram for each circuit and have them checked by the instructor. Wire each circuit in a neat and professional manner and have the installation evaluated by the instructor.
   - Plugging of a three-phase induction motor using a reversing starter and zero-speed switch.
   - Antiplugging of a three-phase induction motor using a reversing starter and zero-speed switch.
   - Dynamic braking of a DC motor.
   - DC injection braking of an induction motor.
   - Electromechanical friction braking of a motor.
4. The purpose of this assignment is to wire and operate different types of motor speed control circuits. Wire as many of the following motor speed control circuits as possible, using whatever type of motor control station is available to you. Complete a line diagram and a wiring diagram for each circuit and have them checked by the instructor. Wire each circuit in a neat and professional manner and have the installation evaluated by the instructor.
   - Multispeed motor starter.
   - Wound-rotor motor controller

9 Motor Control Electronics

PART 1 Quiz: Semiconductor Diodes

Place the answers in the space provided.

1. A diode consists of _______________-type and _______________-type semiconductor materials formed on a single component.
2. The most important operating characteristic of a diode is that it allows current to flow through it in both directions. (True/False) _______________
3. Identify the operating modes of the diodes shown in Figure 9-1.

   a. _______________ bias b. _______________ bias

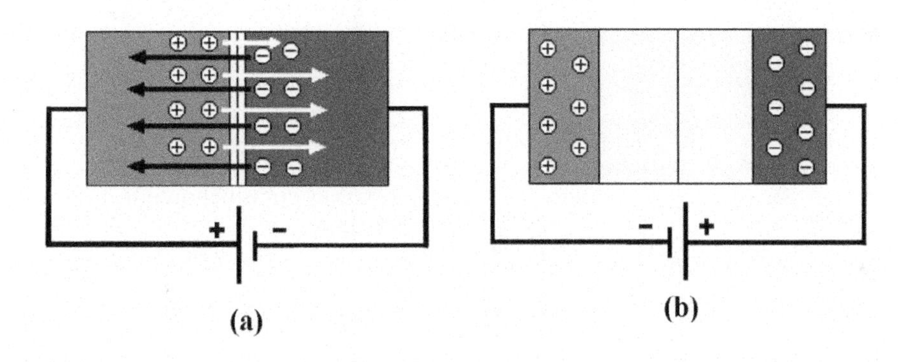

Figure 9-1 Diode for Question 3.

4. Identify the leads of the diode shown in Figure 9-2.

   a. _______________ b. _______________

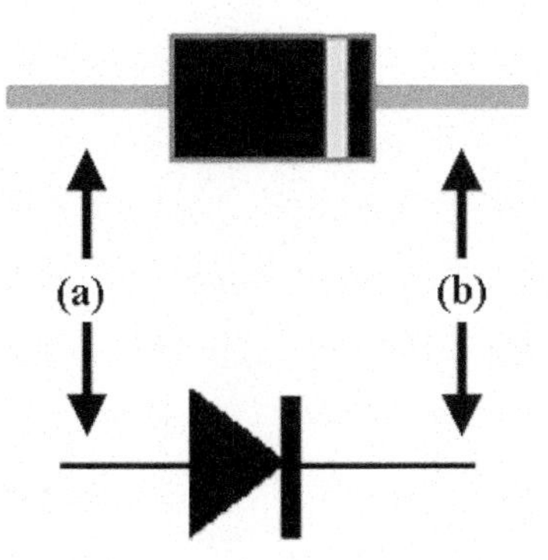

Figure 9-2 Diode for Question 4.

5. When voltage is applied across a diode in such a way that the diode allows current to flow, the diode is said to be
   a. Short-circuited
   b. Open-circuited
   c. Forward-biased
   d. Reverse-biased

   Answer _______

6. Connect the devices shown in Figure 9-3 in series so that the lamp turns on when the switch is closed.

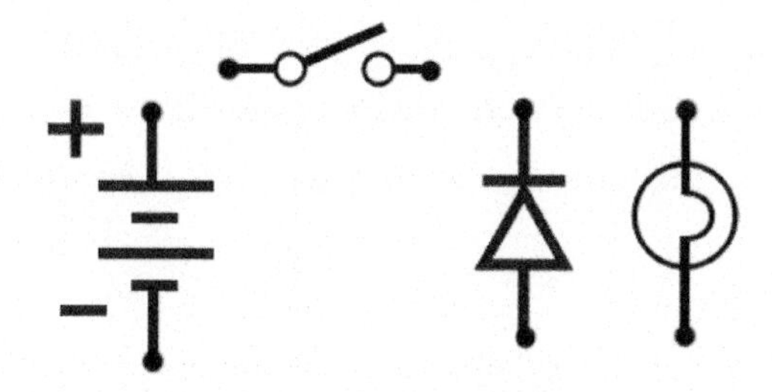

Figure 9-3 Devices for Question 6.

7. Diode rectifiers are used in changing ______________ to ______________.
8. Answer the following with reference to the rectifier circuit shown in Figure 9-4.
   a. This circuit would be classified as a ______________-wave rectifier.
   b. During the positive half-cycle of the AC input, the diode acts as a ______________ switch.
   c. During the negative half-cycle of the AC input, the diode acts as a ______________ switch.
   d. The circuit produces a continuous DC voltage across the load. (True/False) ______________

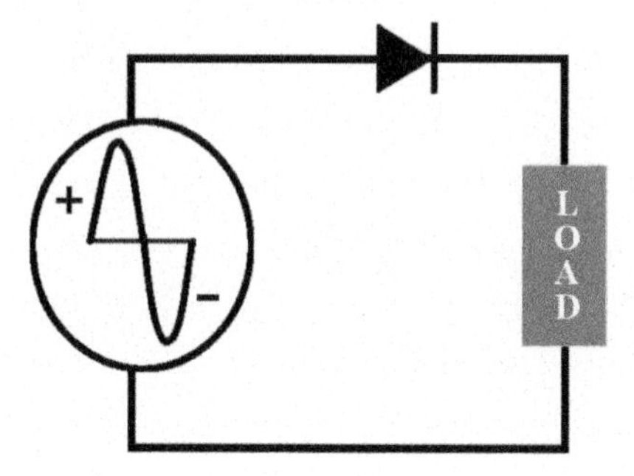

Figure 9-4 Rectifier circuit for Question 8.

9. An ohmmeter test of a diode shows a low resistance reading in both directions. This is an indication that the diode is

   a. Operating normally

   b. Faulted shorted

   c. Faulted open

   d. Faulted grounded

   Answer _______

10. Answer the following with reference to the diode control relay circuit shown in Figure 9-5.

    a. This circuit is an application of a diode-______________ circuit.

    b. The reason the diode is required is because the CR relay coil is mainly a resistive load. (True/False) ______________

    c. Closing the switch results in current flowing through the ______________ but not the ______________.

    d. When the switch is opened, the ______________ voltage of the coil bleeds off through the diode.

    e. Reversing the diode lead connections will result in a short circuit when the switch is closed. (True/False) ______________

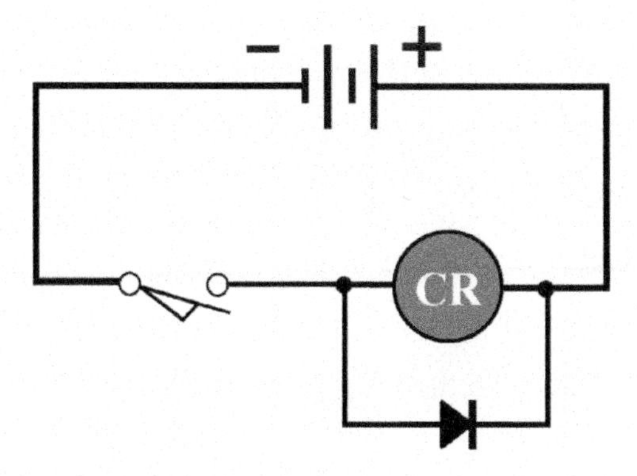

Figure 9-5 Relay circuit for Question 10.

11. Answer the following with reference to the rectifier circuit shown in Figure 9-6.

    a. This circuit would be classified as a _______________-wave rectifier.

    b. On the positive half-cycle, diodes _______________ and _______________ will conduct.

    c. On the negative half-cycle, diodes _______________ and _______________ will conduct.

    d. The shape of the voltage waveform across the load will be similar to the AC input sine wave with the _______________ half of the waveform inverted.

    e. The DC polarity of point A relative to point B will be _______________.

    f. Reversing the two leads of the AC input voltage will result in a reversal of the DC polarity across the load. (True/False) _______________

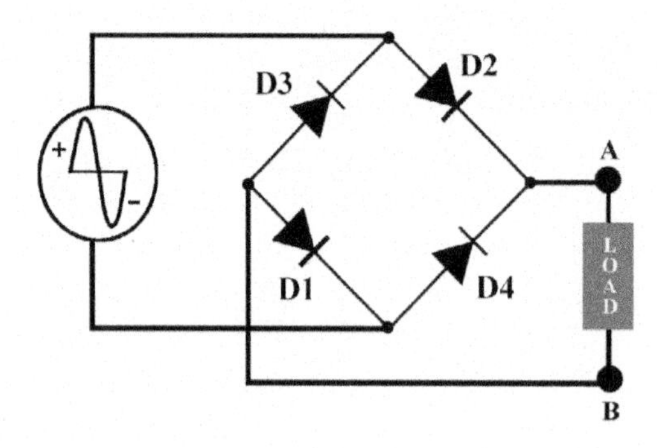

Figure 9-6 Rectifier circuit for Question 11.

12. The purpose of a power supply filter is to

    a. Change alternating current to direct current

    b. Change direct current to alternating current

    c. Smooth out the DC voltage from the diode rectifier circuit

    d. Produce a cleaner form of the original AC input voltage

    Answer _______

13. The DC output voltage of a diode rectifier circuit can be smoothed out by means of a

    a. Resistor connected in series with the output

    b. Resistor connected in parallel with the output

    c. Capacitor connected in series with the output

    d. Capacitor connected in parallel with the output

    Answer _______

14. Answer the following with reference to the rectifier circuit shown in Figure 9-7.
    a. This circuit would be classified as a _______________ full-wave bridge rectifier.
    b. Compared to single-phase rectifiers, this circuit requires much more filtering to obtain a low-ripple DC output. (True/False) _______________
    c. When this rectifier is constructed in modular form, it comes equipped with _______________ AC input leads and _______________ DC output leads.

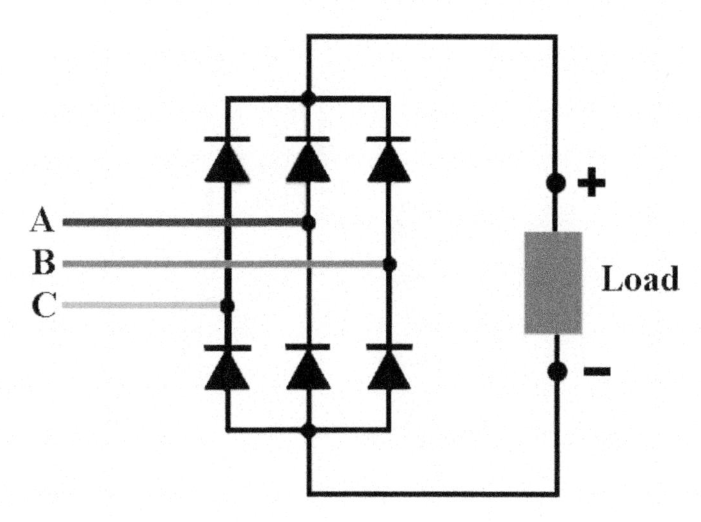

Figure 9-7 Rectifier circuit for Question 14.

15. A zener diode
    a. Allows current to flow in the forward direction
    b. Allows current to flow in the reverse direction when the voltage is above a certain value
    c. Is typically used to regulate voltage in electric circuits
    d. All of these

    Answer _______

16. Answer the following with reference to the circuit shown in Figure 9-8.
    a. The function of this circuit is to suppress damaging AC _______________ voltages.
    b. The varistor module is made up of two _______________ diodes connected back to back.
    c. Each diode acts as an open circuit until the reverse bias voltage across it exceeds its rated value. (True/False) _______________

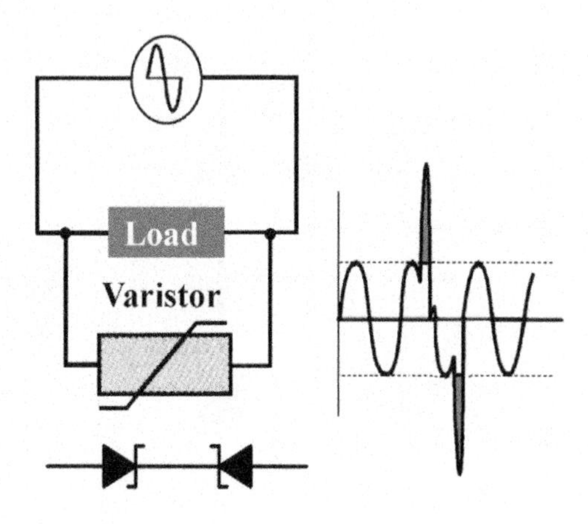

Figure 9-8 Circuit for Question 16.

17. A light-emitting diode (LED) will
    a. Conduct a current in one direction only
    b. Give off light when current flows through it
    c. Consume less energy than a filament-type light bulb
    d. All of these

    Answer _______

18. Answer the following with reference to the optocoupler circuit shown in Figure 9-9.
    a. The diode in the input circuit is a ______________ and is connected in ______________ bias.
    b. The diode in the output circuit is a ______________ and is connected in ______________ bias.
    c. The only thing connecting the two circuits is ______________.
    d. When the push button is closed, ______________ enters the photodiode so its resistance decreases, switching on current to the load.

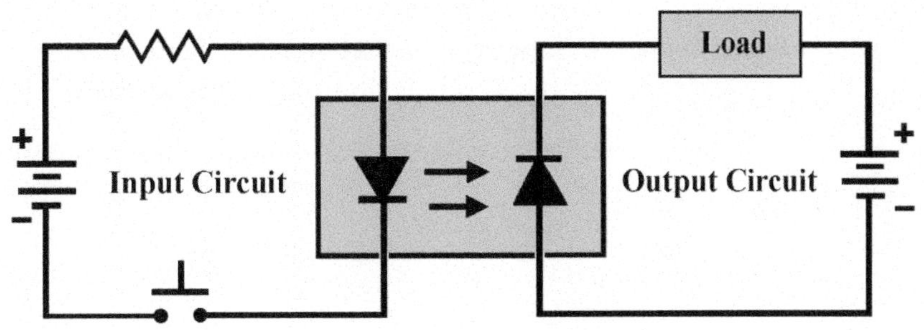

Figure 9-9 Optocoupler circuit for Question 18.

19. An inverter and rectifier perform the same function. (True/False) ______________
20. An inverter is used to convert current from
    a. analog to digital
    b. digital to analog
    c. DC to AC
    d. AC to DC
21. Inverter outputs can be classified as
    a. pure sine wave
    b. modified square wave
    c. square wave
    d. all of these

PART 2 Quiz: Transistors

1. A transistor is used to amplify a signal or switch a circuit on and off. (True/False) ______________
2. The two most common types of transistors are the ______________ transistor and the ______________-______________ transistor.

3. The three leads of a bipolar transistor are the
   a. Anode, base, and cathode
   b. Emitter, gate, and anode
   c. Base, collector, and emitter
   d. Grid, anode, and cathode

   Answer _______
4. In a bipolar transistor circuit, the collector current is typically much higher than the base current. (True/False) ______________
5. The PNP transistor is the complement of the NPN transistor. This means that
   a. The two operate on different principles.
   b. One can be replaced by the other in a given circuit.
   c. The two are electrically similar except that opposite current and voltages are involved.
   d. One is designed to operate from an AC source and the other from a DC source.

   Answer _______
6. Answer the following with reference to the transistor amplifying circuit shown in Figure 9-10.
   a. The bipolar transistor would be classified as an NPN type. (True/False) ______________
   b. The base current varies in direct proportion with the amount of light shining on the photovoltaic sensor. (True/False) ______________
   c. With no light shining on the photovoltaic sensor, the voltage and current generated by it would be ______________.
   d. As the base current increases, the collector current decreases. (True/False) ______________
   e. The current amplification factor or gain for this circuit would be ______________.

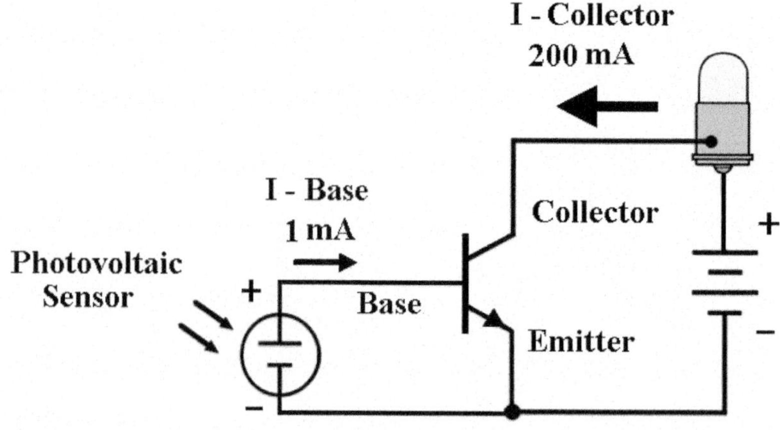

Figure 9-10 Transistor amplifying circuit for Question 6.

7. Answer the following with reference to the transistor switching circuit shown in Figure 9-11.
   a. For the circuit shown the 230-V load is operated directly by the transistor. (True/False) _______________
   b. When the proximity sensor switch is open, the transistor conducts to energize the relay coil. (True/False) _______________
   c. The relay contacts used to control the load would be classified as being normally _______________.
   d. When the transistor is in the off state, the voltage drop across the collector and emitter will be approximately_______________ V and the voltage across the relay coil approximately _______________ V.
   e. The diode used in this circuit prevents the induced coil voltage at turn-off from damaging the transistor. (True/False) _______________

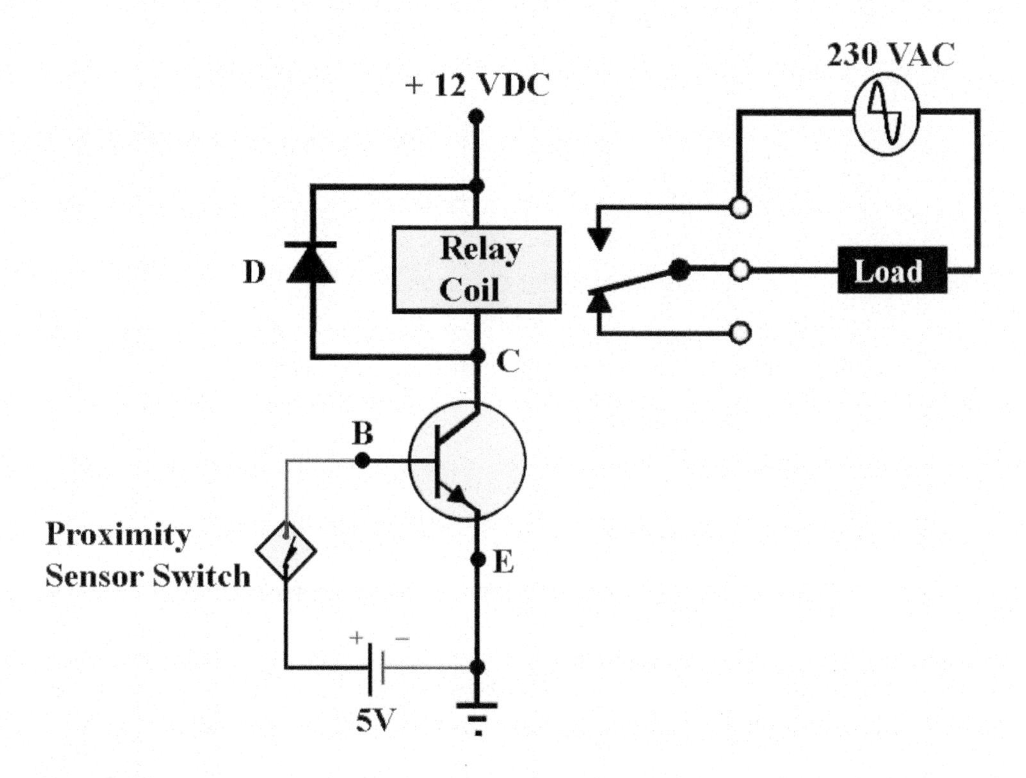

Figure 9-11 Transistor switching circuit for Question 7.

8. A Darlington transistor combines _______________ into a single device.
   a. Two transistors
   b. A transistor and relay
   c. A transistor and diode
   d. A transistor and LED

   Answer ________

9. Identify the labeled components of the phototransistor optical isolator circuit shown in Figure 9-12.
   a. ______________________________
   b. ______________________________
   c. ______________________________
   d. ______________________________

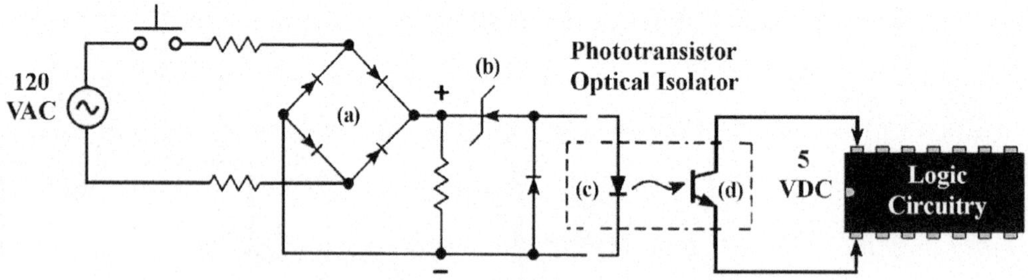

Figure 9-12 Optical isolator circuit for Question 9.

10. Compared to a bipolar transistor, the field-effect transistor (FET)
   a. Has a lower switching speed
   b. Requires less drive power
   c. Is less likely to be damaged by electrostatic discharge
   d. Both a and b

   Answer ________

11. Identify the three leads of the junction field-effect transistor (JFET) shown in Figure 9-13.
   a. ______________
   b. ______________
   c. ______________

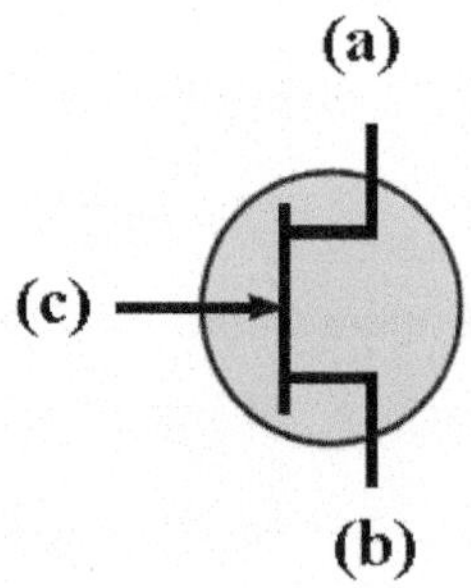

Figure 9-13 Junction field-effect transistor for Question 11.

12. All field-effect transistors are unipolar in that their working current flows through only one type of semiconductor material.

    (True/False) _______________

13. All field-effect transistors use basically no input power, as the output current flow is controlled by an electric field rather than an electric current.

    (True/False) _______________

14. Answer the following with reference to the JFET circuit shown in Figure 9-14.

    a. When the _______________ voltage is zero, maximum current flows between the source and drain.

    b. A negative voltage applied to the gate decreases the channel resistance.

       (True/False) _______________

    c. Passage of current between the source and drain is essentially blocked when the gate voltage reaches the pinch-off point. (True/False) _______________

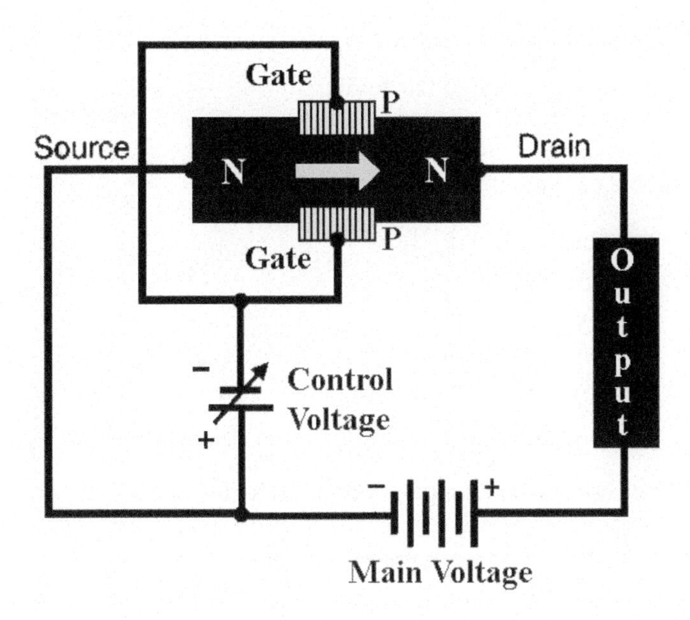

Figure 9-14 JFET circuit for Question 14.

15. The gate of a JFET consists of a _______________-biased junction, whereas the gate of a MOSFET consists of a metal electrode _______________ from the channel.

16. Identify the modes of N-channel MOSFETs represented by the symbols shown in Figure 9-15.

    a. _______________ mode   b. _______________ mode

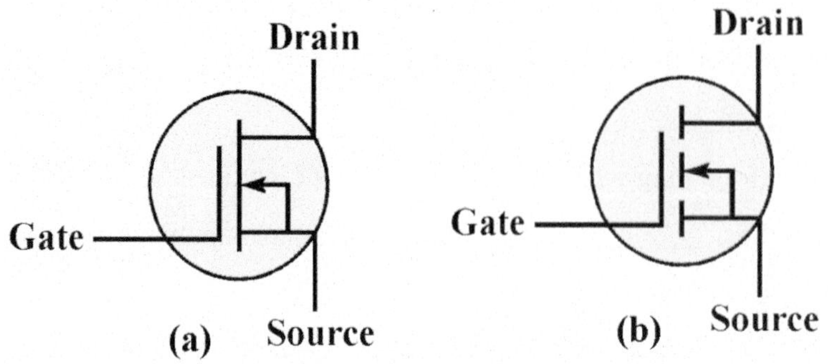

Drain Gate (a) Source   Drain Gate (b) Source

Figure 9-15 N-channel MOSFETs for Question 16.

17. Which of the following field-effect transistors is normally off, in that if a voltage is connected to the drain and source and no voltage is connected to the gate, there will be no current flow through the device?

    a. JFET
    b. Depletion-mode MOSFET
    c. Enhancement-mode MOSFET
    d. Both a and b

    Answer _______

18. Answer the following with reference to the MOSFET off-delay timer circuit shown in Figure 9-16.

    a. With the external switch initially open, the relay coil will be energized. (True/False) _______________
    b. Closing the external switch charges the capacitor and triggers the MOSFET into conduction. (True/False) _______________
    c. When the external switch is closed, the timing period begins. (True/False) _______________
    d. When the external switch is opened, the charged capacitor begins to discharge its stored energy through _______________ and _______________.
    e. If the resistance of the variable resistor R2 is increased, the length of the time-delay period will decrease. (True/False) _______________

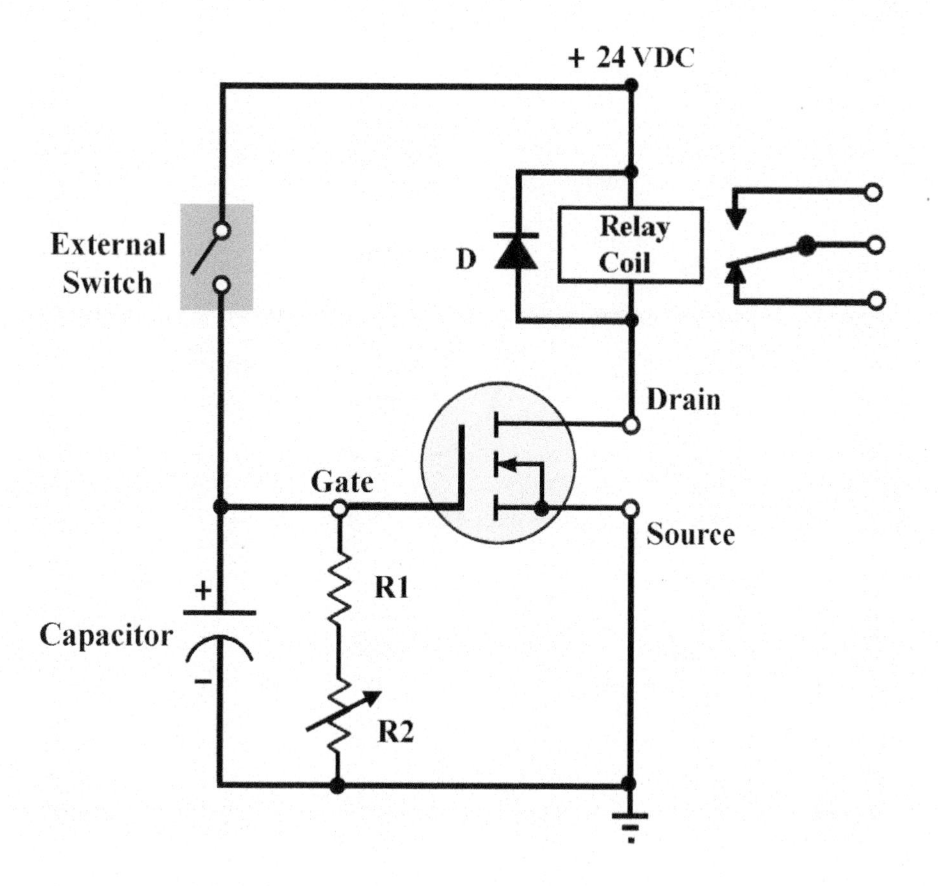

Figure 9-16 MOSFET off-delay timer for Question 18.

19. Compared to mechanical switching circuits, transistor switching circuits
    a. Require less driving power to operate
    b. Can be operated on and off at a higher rate of speed
    c. Have no moving parts to wear out
    d. All of these

    Answer _______

20. Answer the following with reference to the power MOSFET motor control circuit shown in Figure 9-17.
    a. In this application, the MOSFET is used as part of a chopper circuit to switch voltage to the ______________ on and off very rapidly.
    b. By switching the voltage, an average voltage somewhere between 0 and 100 percent is produced. (True/False) ______________

c. When operated with a 50 percent duty cycle, the motor will run slower than when operated with an 80 percent duty cycle. (True/False) ______________

d. The ______________-biased diode provides a discharge path when the motor armature is switched off.

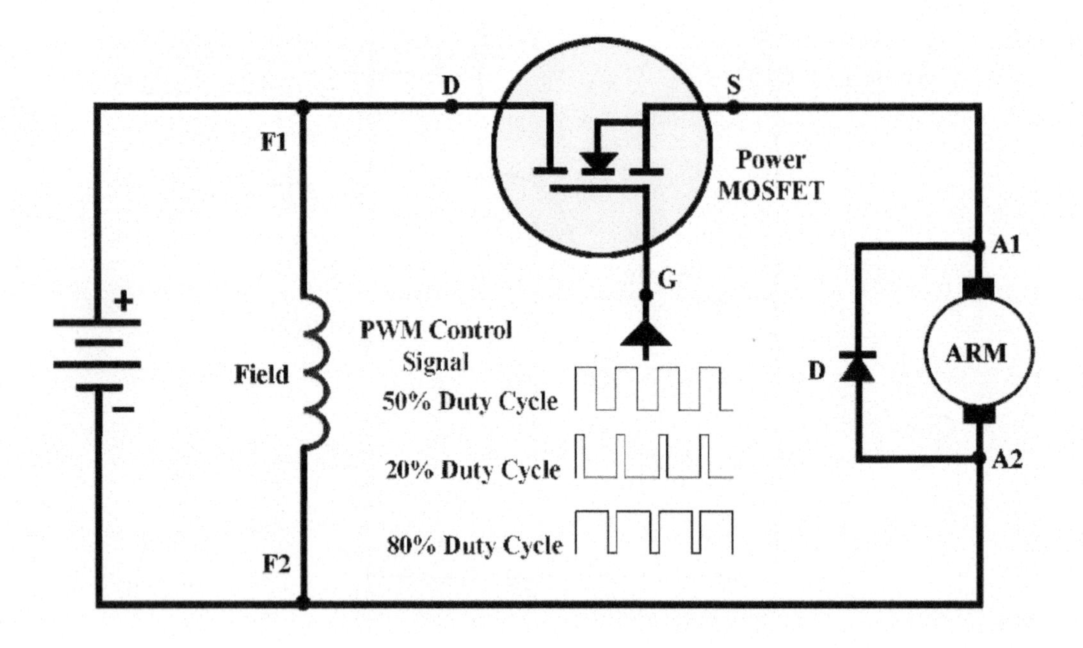

Figure 9-17 MOSFET motor control circuit for Question 20.

21. Identify the leads of the insulated-gate bipolar transistor (IGBT) shown in Figure 9-18.

a. ______________

b. ______________

c. ______________

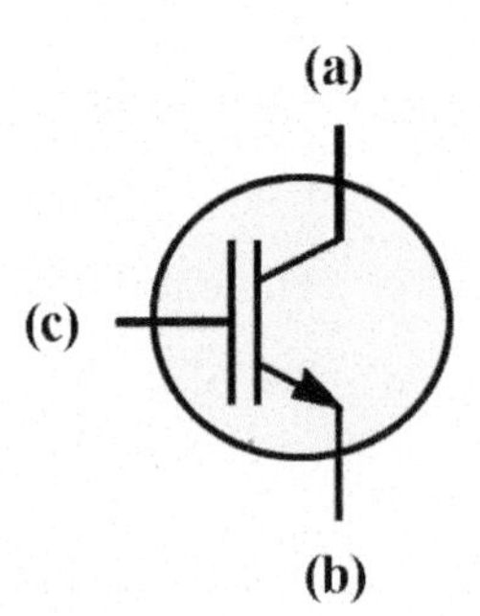

Figure 9-18 IGBT for Question 21.

22. The insulated-gate bipolar transistor is a cross between a ______________ and a ______________ transistor.

23. IGBTs have a lower on-state power loss than MOSFETs. (True/False) ______________

24. IGBTs have faster switching speeds than BJTs. (True/False) ______________

25. Answer the following with reference to the IGBT motor drive circuit shown in Figure 9-19.

   a. The input power from the line is _______________-phase AC.

   b. The six-diode bridge configuration converts the _______________ power to _______________ power.

   c. The inductor and capacitor of the DC bus section act to _______________ out any AC components of the DC waveform.

   d. The six-IGBT bridge configuration inverts the _______________ power back to _______________ power.

   e. Control circuit switching of the IGBT devices delivers a variable _______________ and _______________ output to the motor.

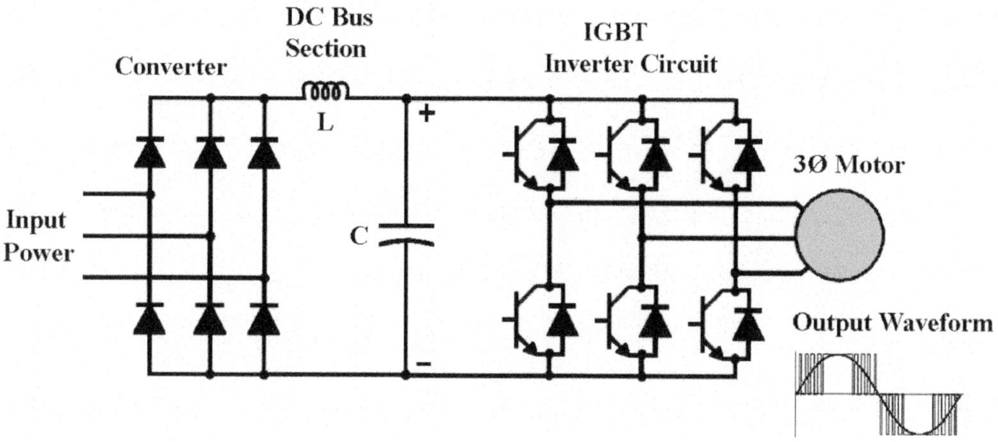

Figure 9-19 IGBT motor drive circuit for Question 25.

26. The function of a driver in an electronic motor control system is to provide signals/information to the controller.
(True/False) _______________

PART 3 Quiz: Thyristors

1. Thyristors have only two states—_______________ and _______________.
2. Thyristors are mainly used where low currents and voltages are involved.
(True/False) _______________

3. Identify the leads of the silicon-controlled rectifier (SCR) shown in Figure 9-20.

   a. _______________

   b. _______________

   c. _______________

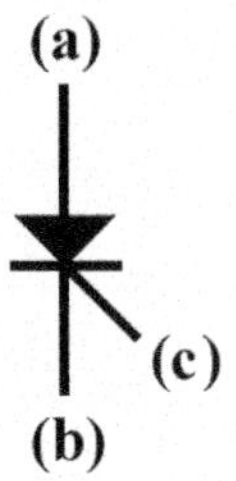

(a)

(c)

(b)

Figure 9-20
SCR for Question 3.

4. High-current SCRs have provisions for some type of heat _______________ to dissipate the heat.

5. How is a silicon-controlled rectifier similar to a diode rectifier?

   a. Both conduct current in one direction only.

   b. Both can be classified as thyristors.

   c. Both are constructed using only one PN junction.

   d. All of these.

   Answer _______

6. Answer the following with reference to the direct-current SCR switching circuit shown in Figure 9-21.

   a. Momentarily pressing pushbutton _______________ switches the SCR into conduction.

   b. Once switched on, the SCR conducts like a _______________.

   c. Once conducting, the only way to turn the SCR off is to reduce the _______________-to-_______________ current to zero.

   d. Reversing the polarity of the DC supply voltage will reverse which push buttons turn the lamp on and off. (True/False) _______________

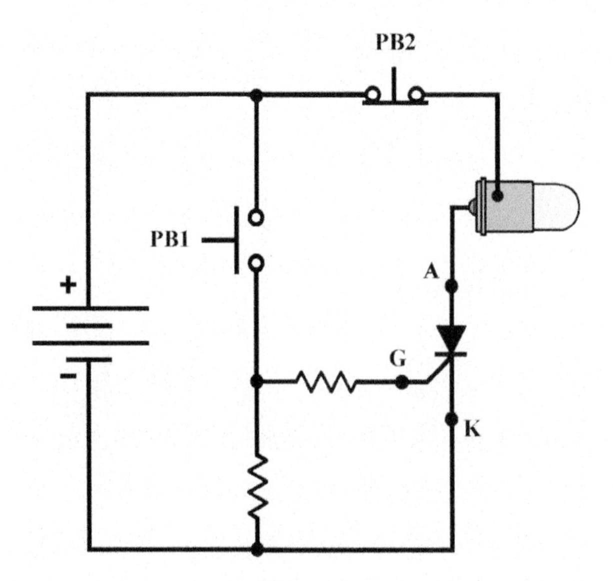

Figure 9-21 Direct current SCR switching circuit for Question 6.

7. Answer the following with reference to the alternating current SCR switching circuit shown in Figure 9-22.
   a. Momentarily closing PB will switch the lamp on, and it will remain on after the push button returns to its normally open position.
      (True/False) _______________
   b. The SCR is automatically shut off during each _______________ half-cycle because it is _______________-biased.
   c. Holding PB closed will produce a _______________-wave pulsating direct current through the lamp.

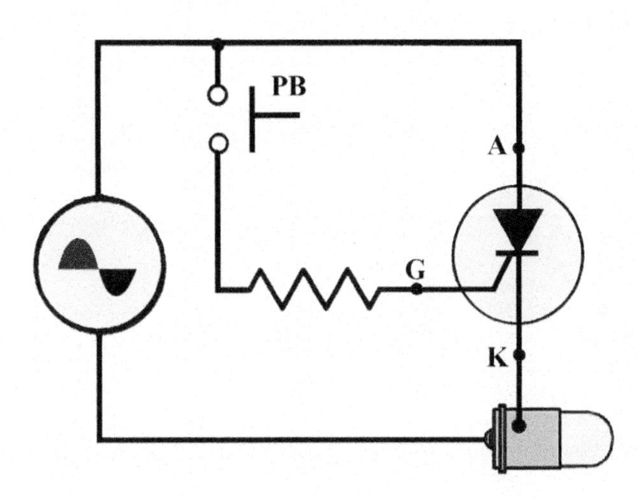

Figure 9-22 Alternating current SCR switching circuit for Question 7.

8. When an SCR is operated from an AC source
   a. The gate can be used to switch the current on and off at all times
   b. Once switched on by the gate, the current cannot be switched off by the gate
   c. The output is always direct current
   d. Both b and c

   Answer _______
9. Answer the following with reference to the alternating current SCR circuit shown in Figure 9-23 used to vary the amount of power delivered to a load.
   a. The load voltage is varied by using phase shift control.
      (True/False) _______________
   b. A pulse trigger is applied to the _______________ lead at the instant that the SCR is required to turn on.

c. During the positive half of the AC input waveform, SCR-_______________ and SCR-_______________ can be triggered into conduction.

d. During the negative half of the AC input waveform, SCR-_______________ and SCR-_______________ can be triggered into conduction.

e. Power is regulated by advancing or delaying the point at which each pair of SCRs is turned on within each half cycle. (True/False) _______________

f. Applying the trigger pulse at the midpoint of the half-cycle will result in the delivery of maximum power to the load. (True/False) _______________

g. The current through the load reverses as the AC source voltage alternates from one half-cycle to the other. (True/False) _______________

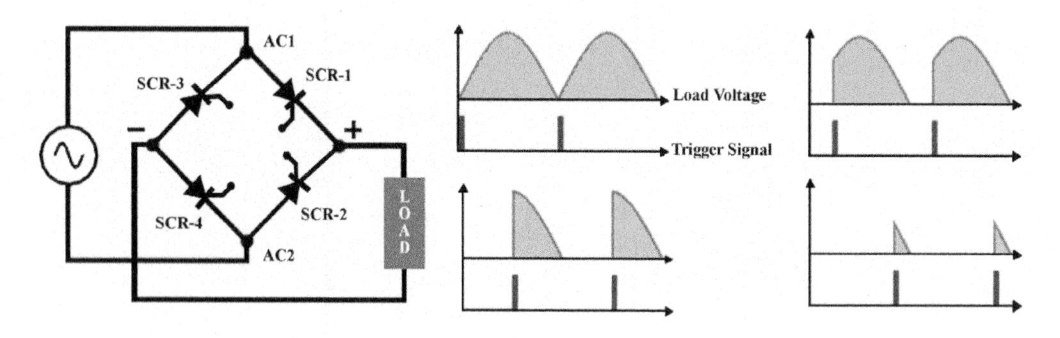

Figure 9-23 Alternating current SCR circuit for Question 9.

10. Answer the following with reference to the SCR solid-state reduced-voltage starter circuit shown in Figure 9-24.

a. In this application, a controlled AC output is obtained by connecting two SCRs in _______________-parallel.

b. When the motor is first started, contacts _______________ close to apply a reduced voltage to the motor.

c. Controlled triggering of the SCRs allows portions of the AC input waveform to be applied to the motor. (True/False) _______________

d. The current transformers are used to monitor motor speed. (True/False) _______________

e. After the starting period, contacts _______________ open and contacts _______________ connect the motor directly across the line.

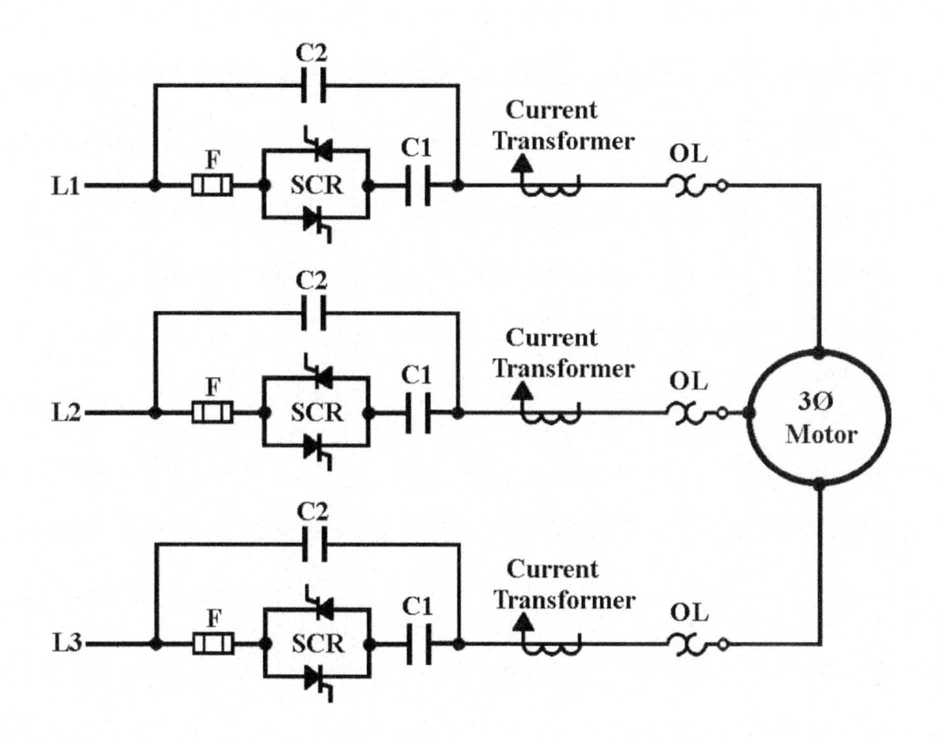

Figure 9-24 SCR solid-state reduced-voltage starter circuit for Question 10.

11. SCRs usually fail shorted as indicated by an ohmmeter resistance reading from anode to cathode that is

    a. High in one direction and low in the other

    b. High in both directions

    c. Near infinity in both directions

    d. Near zero in both directions

    Answer _______

12. The physical size of an SCR has no relationship to its current rating.

    (True/False) _______________

13. Identify the leads of the triac shown in Figure 9-25.

    a. _______________

    b. _______________

    c. _______________

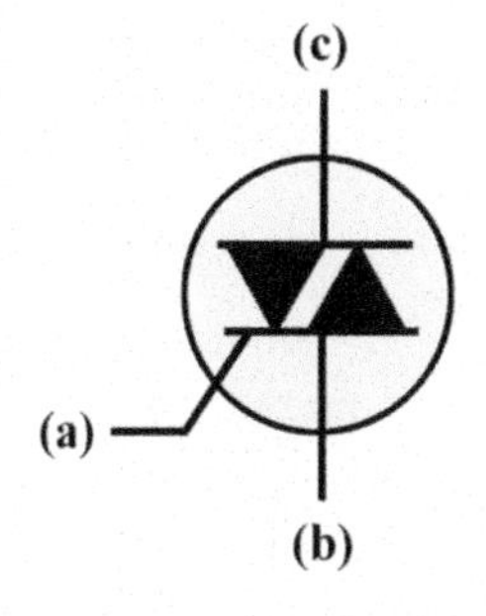

Figure 9-25 Triac for Question 13.

14. Current flow between the main terminals of a triac is
    a. In one direction only
    b. In either direction
    c. Constant at all times
    d. Less than the gate current flow value

    Answer _______

15. Answer the following with reference to the triac circuit shown in Figure 9-26.
    a. This circuit is used to switch DC loads. (True/False) ______________
    b. When the switch is closed and then opened, the triac will conduct and remain conducting. (True/False) ______________
    c. The function of the resistor is to limit the flow in the gate circuit to a small control current value. (True/False) ______________
    d. The current rating of the switch can be much lower than that of the load. (True/False) ______________

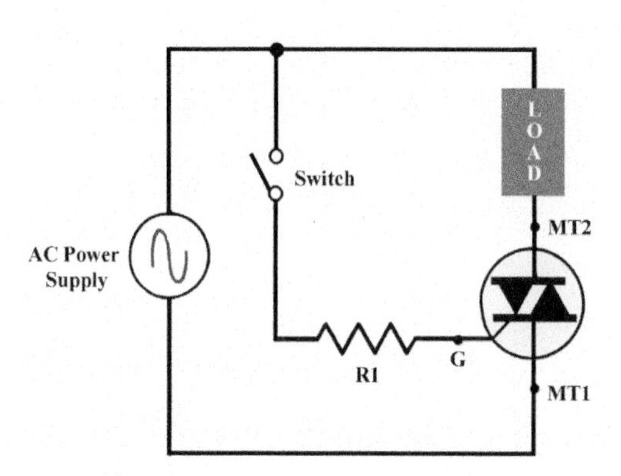

Figure 9-26 Triac circuit for Question 15.

16. A triac can serve as the switching device for a programmable logic controller AC output module. (True/False) ______________

17. Answer the following with reference to the triac variable AC control circuit shown in Figure 9-27.
    a. The control logic controls the ______________ on the AC waveform at which the triac is switched on.
    b. The output waveform is still AC, but the ______________ current value is adjustable.

c. This circuit could be used to vary the speed of a single-phase universal motor. (True/False) ______________

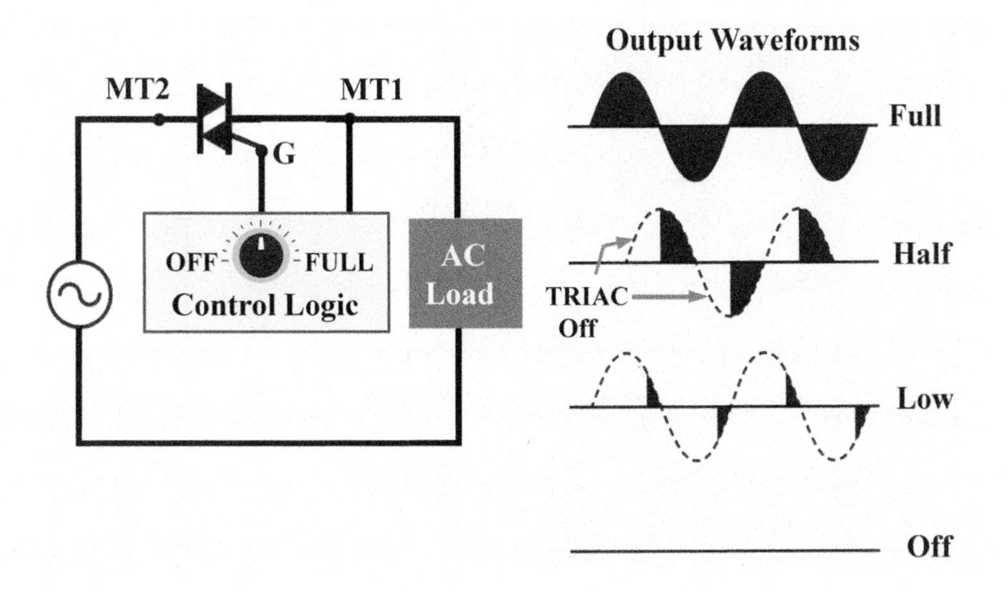

Figure 9-27 Triac AC control circuit for Question 17.

18. A diac is designed to conduct current in
    a. Both directions like a resistor
    b. The forward-bias direction only
    c. The reverse-bias direction only
    d. Either direction when its rated brakeover voltage is exceeded

    Answer _______
19. Answer the following with reference to the lamp dimmer circuit shown in Figure 9-28.
    a. Maximum lamp brightness is achieved when the variable resistor is set to its maximum resistance value. (True/False) ______________
    b. The diac conducts at all times that the capacitor is charging. (True/False) ______________
    c. If the triac is faulted shorted, the lamp will be at maximum brightness at all times. (True/False) ______________

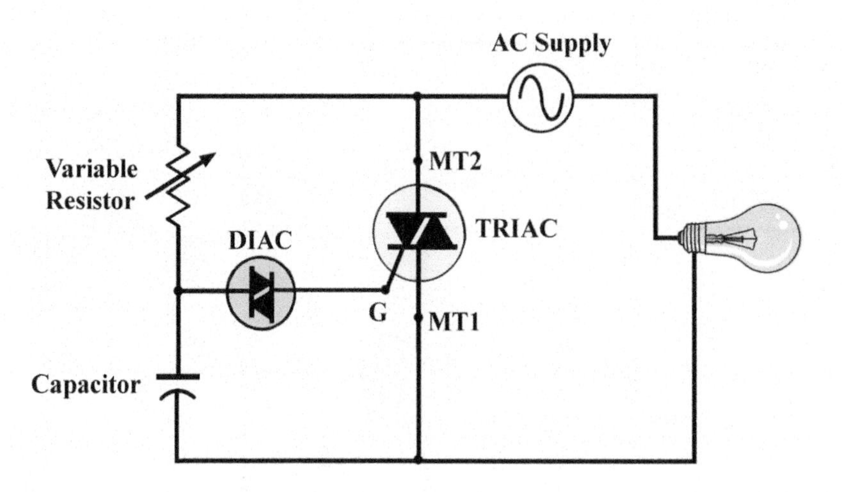

Figure 9-28 Lamp dimmer circuit for Question 19.

20. Which of the following building blocks of an electronic motor control system takes command of motor starting, stopping, speed, and protection?
    a. Controller
    b. Driver
    c. Sensors
    d. Rheostat

PART 4 Quiz: Integrated Circuits (ICs)

1. An integrated circuit (IC) is a
    a. Number of transistors mounted within a silicon chip
    b. Miniaturized electronic circuit fabricated into a single silicon chip
    c. Series of discrete components wired together
    d. Series of separately produced components wired within a printed circuit board

    Answer _______
2. ICs are favored for high-voltage and high-current applications. (True/False) _______________
3. Digital ICs operate with on/off _______________-type signals.
4. Analog ICs contain _______________-type circuitry.

5. Identify the two pin numbers of the IC shown in Figure 9-29.

   a. ____________ b. ____________

Figure 9-29 IC for Question 5.

6. An operational amplifier (op-amp) would be classified as

   a. An analog IC
   b. A digital IC
   c. A MOS IC
   d. A TTL IC

   Answer _______

7. Answer the following with reference to the op-amp voltage-amplifier circuit shown in Figure 9-30.

   a. Resistor ______________ is called the feedback resistor.
   b. The voltage gain of this circuit would be ______________.
   c. When the peak-to-peak value of the input signal is 100 mV, the peak-to-peak value of the output signal would be ______________.
   d. The output signal is in phase with the input signal. (True/False) ______________

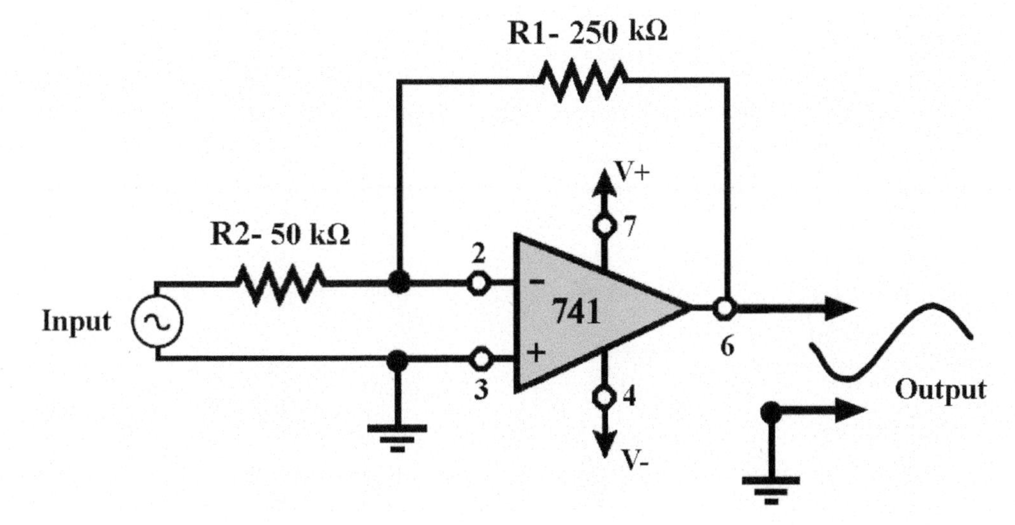

Figure 9-30 Op-amp circuit for Question 7.

8. Answer the following with reference to the op-amp voltage comparator circuit shown in Figure 9-31.

   a. The output is dependent on the difference in value between voltages V1 and V2. (True/False) _______________

   b. When the light-dependent resistor (LDR) is not illuminated with light, its resistance is very high. (True/False) _______________

   c. When sufficient light shines on the LDR, voltage V1 will be greater than V2, resulting in an output from the op-amp activating the relay to switch the connected load. (True/False) _______________

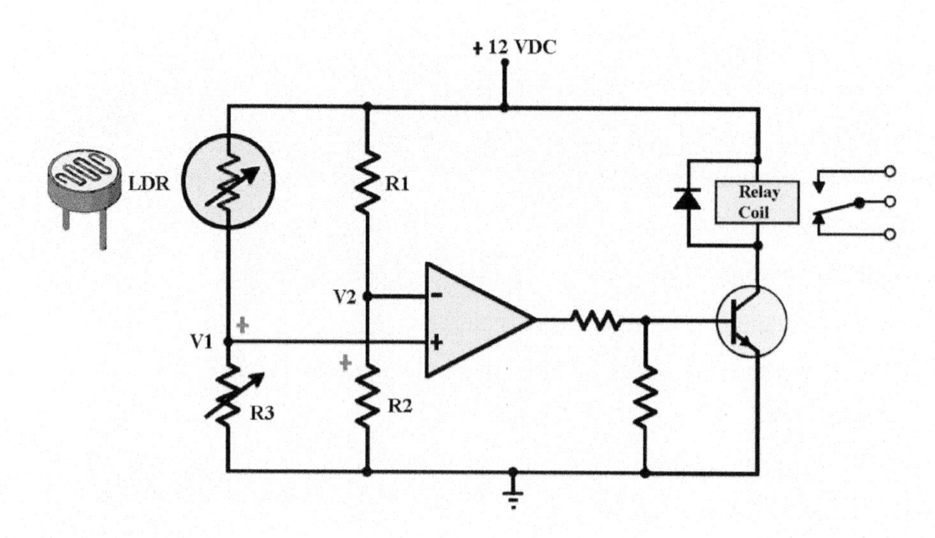

Figure 9-31 Op-amp voltage comparator circuit for Question 8.

9. Answer the following with reference to the 555 IC interval timer circuit shown in Figure 9-32.

   a. The length of the time interval is determined by the value of timing components _______________ and _______________.

   b. When the switch is opened, the capacitor is held charged. (True/False) _______________

   c. When the switch is closed, the capacitor starts to charge. (True/False) _______________

   d. When the switch is closed, the LED is immediately switched on. (True/False) _______________

   e. When the switch is opened, the LED is immediately switched off. (True/False) _______________

f. With the switch kept closed, the LED remains on until the charge on the capacitor reaches a certain level. (True/False) _______________

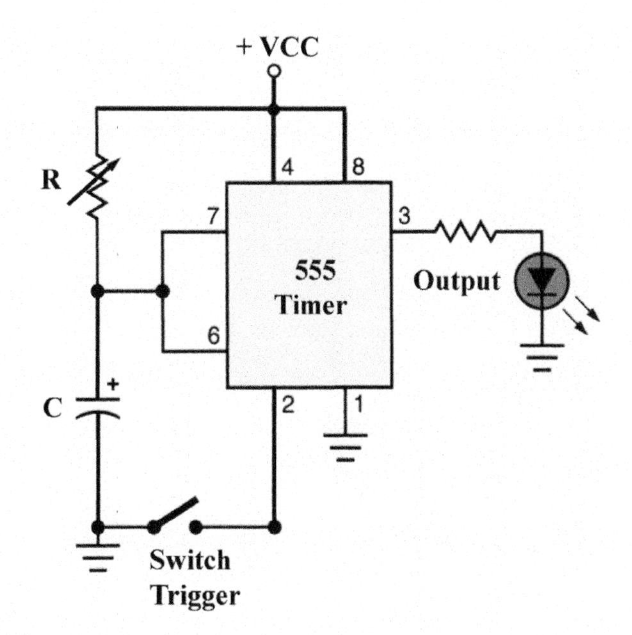

Figure 9-32 IC interval timer circuit for Question 9.

10. Answer the following with reference to the 555 DC motor speed controller circuit shown in Figure 9-33.

a. The type of speed control used in this circuit is known as _______________ _______________ _______________.

b. The voltage applied to the motor armature is switched on and off by transistor Q2. (True/False) _______________

c. A duty cycle of 50 percent will result in a greater armature voltage and speed than one of 10 percent. (True/False) _______________

d. Potentiometer R1 controls the length of time the _______________ of the timer will be turned on.

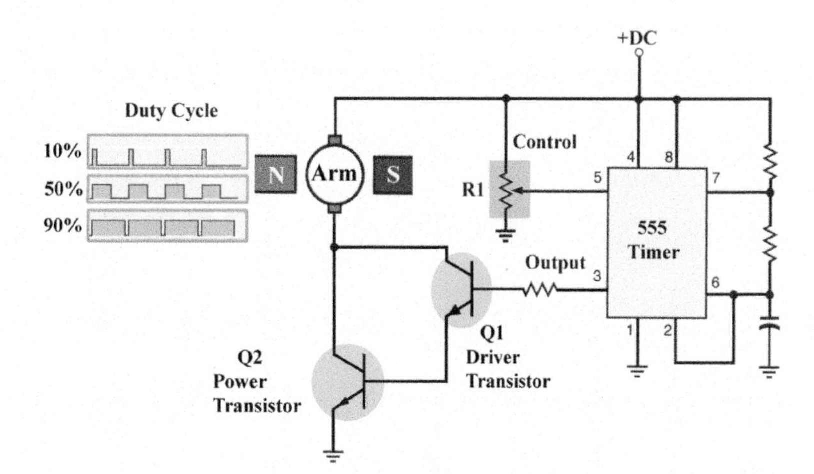

Figure 9-33 Speed control circuit for Question 10.

11. A microcontroller is an integrated circuit that functions as a complete computer on a chip. (True/False) ______________

12. An ______________ microcontroller is one that is physically built into the device it controls.

13. Electrostatic discharge (ESD) is the release of static electricity when two objects come into contact. (True/False) ______________

14. The resultant ______________ flow from ESD can generate damaging heat within an IC.

15. Always transport and store sensitive ICs and control boards in ______________ packaging.

16. Logic is the ability to make decisions based on one or more different factors before an action is taken. (True/False) ______________

17. The only two states in digital logic circuits are ______________ and ______________.

18. Digital circuits are constructed from electronic circuits known as logic ______________.

19. Identify the type of logic gate function represented by the symbols shown in Figure 9-34.

    a. ______________ b. ______________ c. ______________

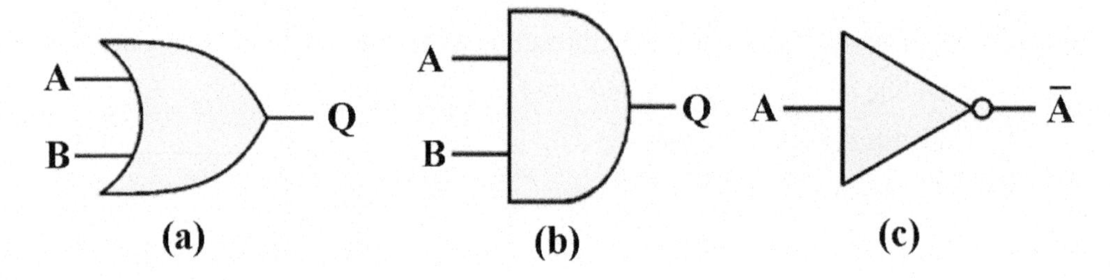

Figure 9-34 Symbols for Question 19.

20. A logic gate ______________ table shows the output for each possible input.

21. The table shown in Figure 9-35 summarizes the operation of a two-input
    a. AND gate
    b. OR gate
    c. NOT gate
    d. NAND gate

    Answer _______

Input A	Input B	Output Q
0	0	0
0	1	1
1	0	1
1	1	1

Figure 9-35 Table for Question 21.

22. The table shown in Figure 9-36 summarizes the operation of a two-input
    a. AND gate
    b. OR gate
    c. NOT gate
    d. NAND gate

    Answer _______

Input A	Input B	Output Q
0	0	1
0	1	1
1	0	1
1	1	0

Figure 9-36 Table for Question 22.

23. Identify the type of gate logic function utilized in the hard-wired control circuit shown in Figure 9-37.

    Answer _______

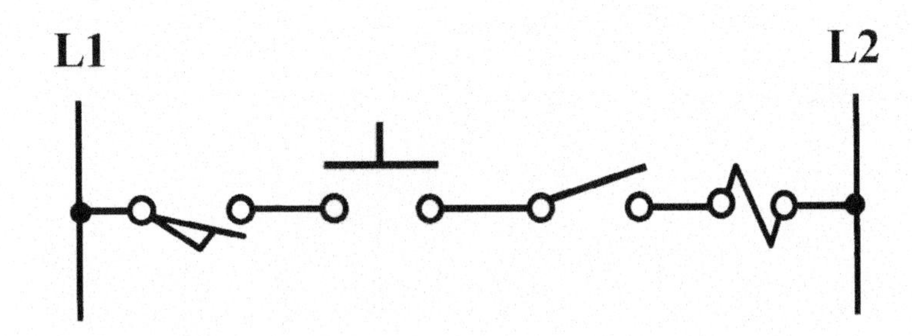

Figure 9-37 Control circuit for Question 23.

24. Identify the type of gate logic function utilized in the hard-wired control circuit shown in Figure 9-38.

Answer _______

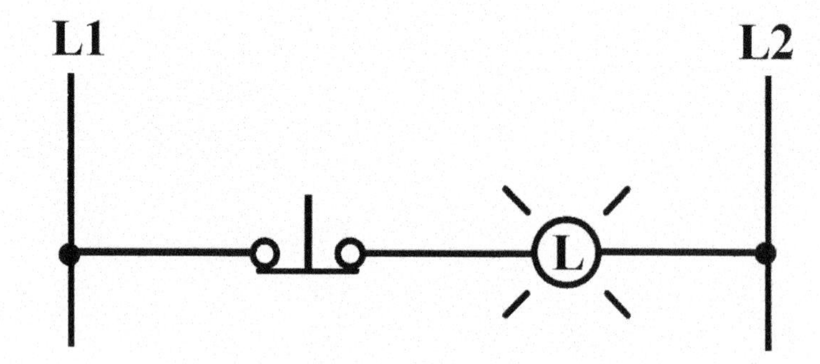

Figure 9-38 Control circuit for Question 24.

25. Identify the type of gate logic function utilized in the programmable logic controller (PLC) circuit shown in Figure 9-39.

Answer _______

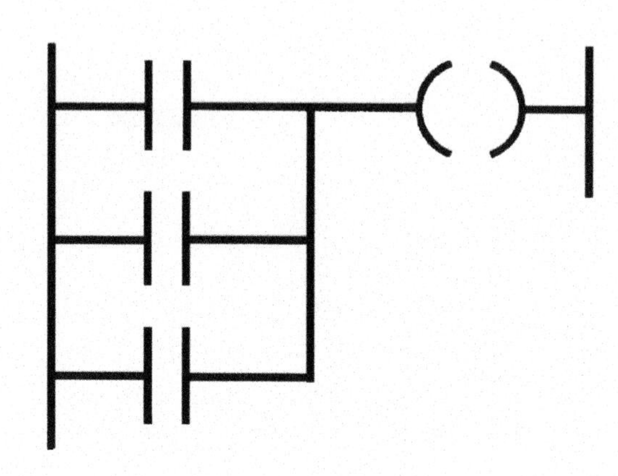

Figure 9-39 PLC circuit for Question 25.

26. Which of the following types of digital signals provides a single piece of information?

a. Nibble
b. Byte
c. Word
d. Bit

27. Which type of three-wire sensor is shown?

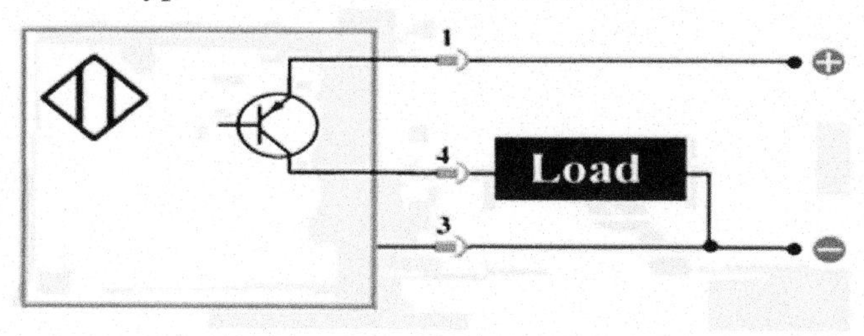

a. Sourcing
b. Sinking
c. NPN
d. SCR

10 Adjustable-Speed Drives and PLC Installations

PART 1 Quiz: AC Motor Drive Fundamentals

Place the answers in the space provided.

1. An adjustable-speed drive may be used to control _______________ of a motor
   a. Speed and torque
   b. Acceleration and deceleration
   c. Direction of rotation
   d. All of these

   Answer _______
2. The preferred method of speed control for squirrel-cage induction motors is to alter the _______________ of the supply voltage.
3. Identify the blocks of the block diagram of the variable-frequency drive (VFD) shown in Figure 10-1.

   a. _______________ b. _______________

   c. _______________ d. _______________

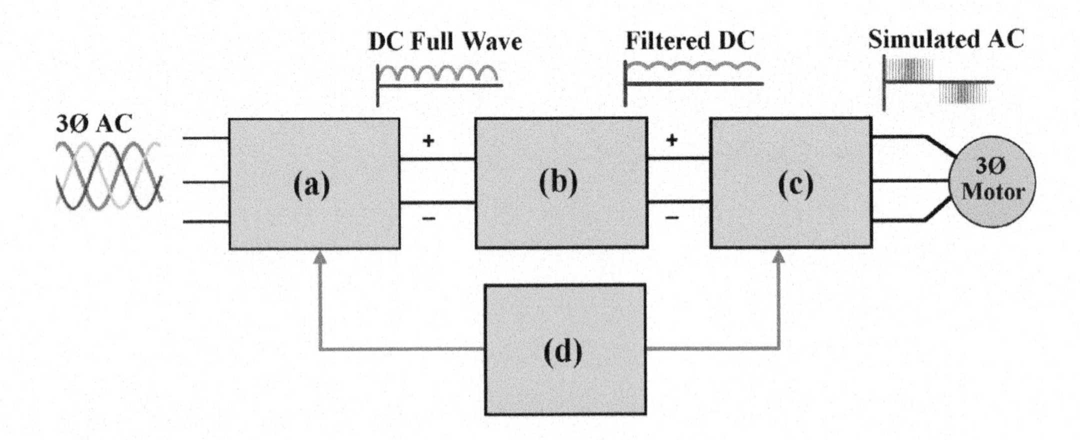

Figure 10-1 Variable-frequency drive for Question 3.

4. A VFD can be used to operate a three-phase motor from a single-phase AC supply. (True/False) _______________

5. Complete the wiring for the three-phase VFD converter circuit shown in Figure 10-2.

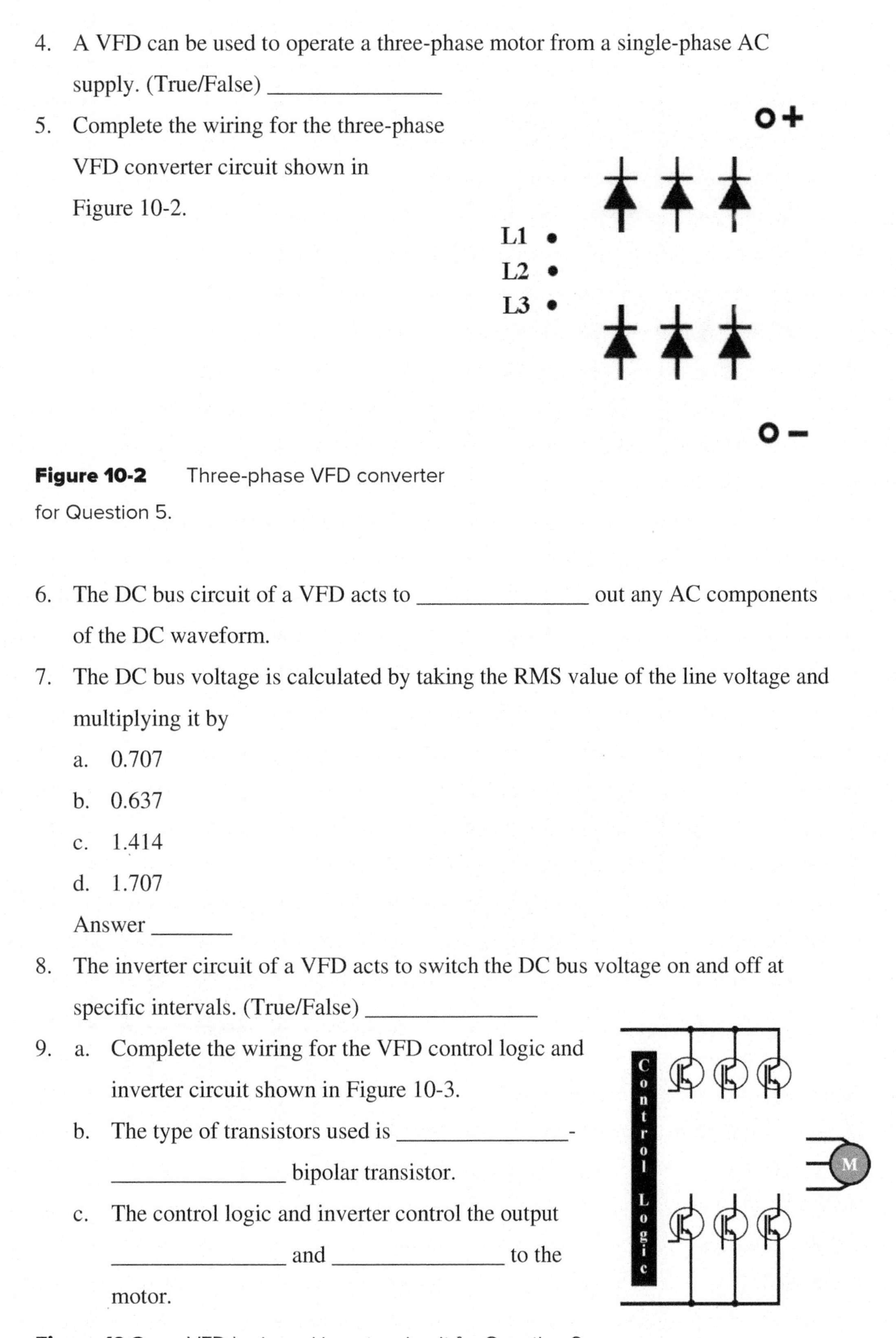

Figure 10-2 Three-phase VFD converter for Question 5.

6. The DC bus circuit of a VFD acts to _______________ out any AC components of the DC waveform.

7. The DC bus voltage is calculated by taking the RMS value of the line voltage and multiplying it by
   a. 0.707
   b. 0.637
   c. 1.414
   d. 1.707

   Answer _______

8. The inverter circuit of a VFD acts to switch the DC bus voltage on and off at specific intervals. (True/False) _______________

9. a. Complete the wiring for the VFD control logic and inverter circuit shown in Figure 10-3.
   b. The type of transistors used is _______________-_______________ bipolar transistor.
   c. The control logic and inverter control the output _______________ and _______________ to the motor.

Figure 10-3 VFD logic and inverter circuit for Question 9.

10. Answer the following with reference to the inverter voltage output waveforms shown in Figure 10-4.

   a. This inverter would be classified as a pulse-width modulated type. (True/False) _______________

   b. The sine wave represents the _______________ frequency and the pulses the _______________ frequency.

   c. The pulses may be of different heights. (True/False) _______________

   d. The output voltage is varied by changing the _______________ and _______________ of the switched pulses.

   e. The output _______________ is adjusted by changing the switching cycle time.

   f. The fundamental frequency is a fixed frequency substantially higher than the switching frequency. (True/False) _______________

   g. The whining sound associated with VFDs is generated by the fundamental frequency. (True/False) _______________

   h. Higher carrier frequency allows a better approximation to the sinusoidal form of the output current. (True/False) _______________

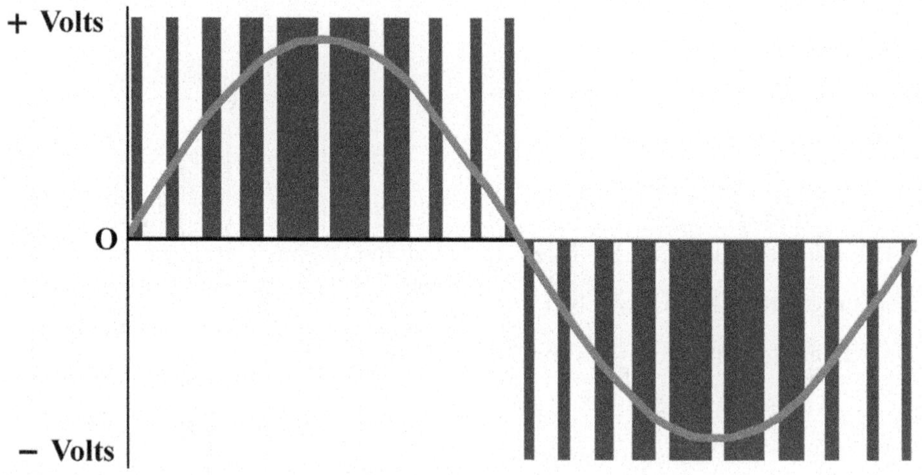

+ Volts

O

– Volts

Figure 10-4 Waveforms for Question 10.

11. The _______________-duty motor is designed for optimized performance to operate in conjunction with a VFD.

12. An open-loop VFD operates without benefit of any _______________ signals.

13. With VFD speed control the voltage to the motor must be increased as the frequency is decreased to prevent overheating of the motor. (True/False) _______________

14. Answer the following with reference to the VFD voltage versus frequency (V/Hz) graph shown in Figure 10-5.

    a. The ratio between voltage and frequency does not change as the motor voltage is increased from 0 to 460 V. (True/False) _______________

    b. When 230 V is applied to the motor, the frequency at which this voltage is applied will be approximately _______________ Hz.

    c. As the voltage to the motor increases, the motor speed decreases. (True/False) _______________

    d. As the voltage increases from zero to the base speed 460-V level, the _______________ remains constant.

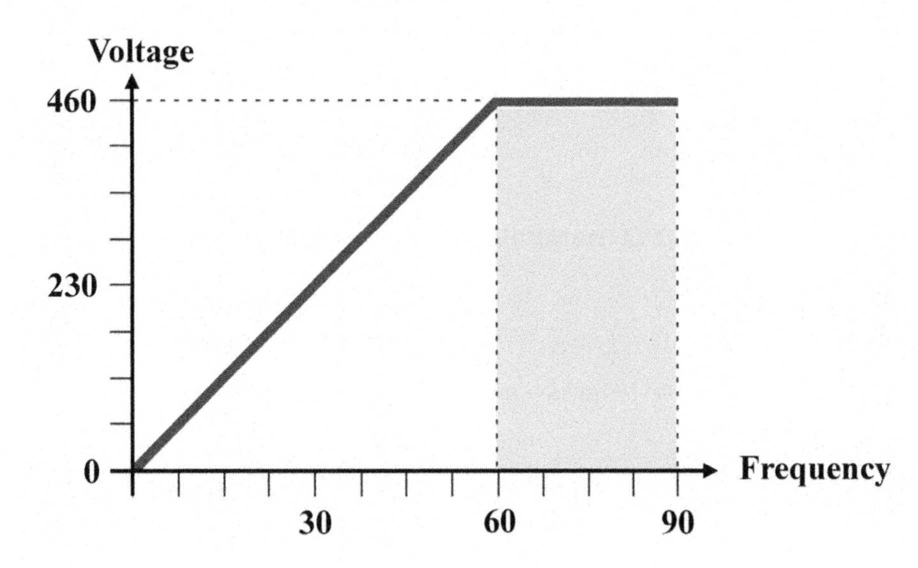

Figure 10-5 Voltage versus frequency graph for Question 14.

15. Which of the following VFD refinements compensates for power losses that occur at low speeds?

    a. Stability compensation

    b. Steady-state slip compensation

    c. Low-frequency voltage boost

    d. Both a and b

    Answer _______

16. The flux vector method of VFD speed control provides less precise control of motor speed and torque. (True/False) _______________

17. Flux vector VFDs are classified as _______________ types.

    a. Sensor

    b. Sensorless

    c. Magnetic

    d. Either a or b

    Answer _______

18. A motor-attached encoder device may be used to provide feedback on the motor's

    a. Rotor position

    b. Speed

    c. Direction

    d. All of these

    Answer _______

PART 2 Quiz: VFD Installation and Programming Parameters

1. The load characteristics of the driven machinery are not of importance when you are selecting a variable-frequency drive for a particular application. (True/False) _______________

2. A VFD reactor is basically an _______________ installed on the input or output of the drive.

3. Use of which of the following will increase the distance that the motor can be located from the drive?

    a. Circuit breaker protection

    b. Fuse protection

    c. Load reactor

    d. Line reactor

    Answer _______

4. Which of the following should be taken into consideration in determining the location for a VFD?
   a. Distance between the motor and the drive
   b. Surrounding environment
   c. Working space to carry out maintenance
   d. All of these

   Answer ________
5. VFD ________________ are designed to provide protection from the environment.
6. Answer the following with reference to the VFD operator interface shown in Figure 10-6.
   a. The three modes of operation are ________________, ________________, and ________________.
   b. The interface provides a means for the operator to ________________ and ________________ the motor.
   c. List five motor typical operating parameters that may be viewed in real time on the monitor.

      (1) ________________ (2) ________________

      (3) ________________ (4) ________________

      (5) ________________

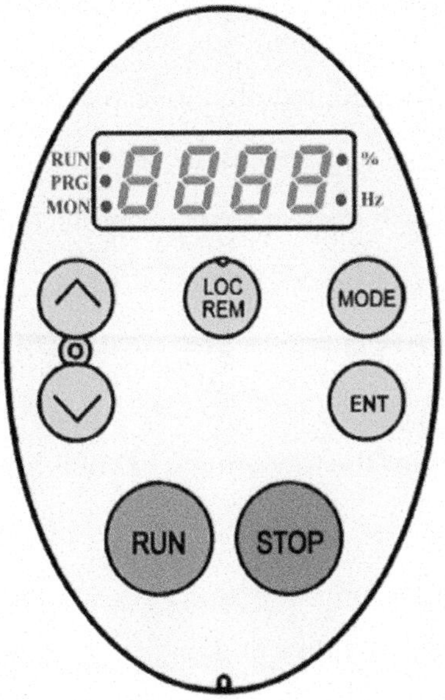

Figure 10-6 VFD operator interface for Question 6.

7. Electromagnetic interference (EMI) is generated in a VDF as a result of the high switching ________________ of the transistors.

8. Most manufacturers require that a shielded power cable be used for the electrical connection between the VFD and the motor. (True/False) _______________

9. In addition to the safety aspect, proper grounding of a VFD system reduces electromagnetic interference. (True/False) _______________

10. For the bypass contactor power circuit shown in Figure 10-7, when the motor is required to be directly operated from the line, the

    a. Isolation contacts are opened and the bypass contacts closed

    b. Isolation contacts are closed and the bypass contacts open

    c. Both the isolation and bypass contacts are opened

    d. Both the isolation and bypass contacts are closed

    Answer _______

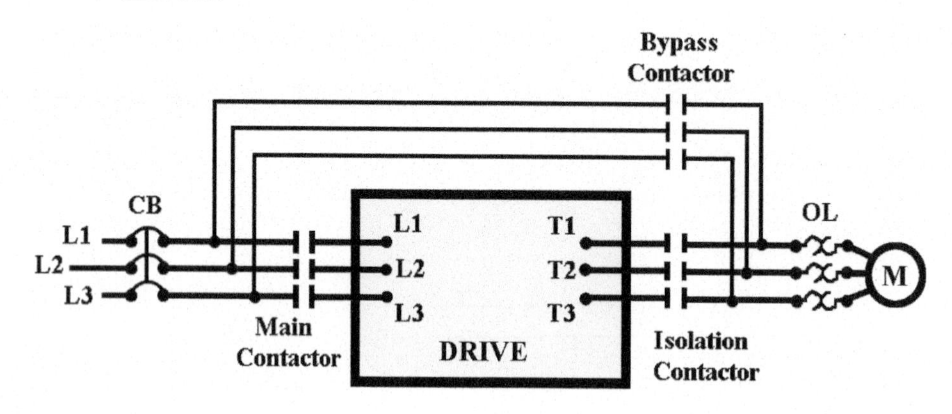

Figure 10-7 Bypass contactor circuit for Question 10.

11. VFDs are not required to have a disconnecting means capable of disconnecting all power supplied to the drive. (True/False) _______________

12. VFDs can operate as motor protection devices along with their role of motor speed controllers. (True/False) _______________

13. Excessive energy is generated when the _______________ drives the motor during deceleration.

14. Which of the following VFD braking systems redirects energy back to the AC supply?

    a. Dynamic braking

    b. Regenerative braking

    c. Friction braking

    d. Both a and b

    Answer _______

15. _______________ is the ability of a VFD to increase or decrease the voltage and frequency to an AC motor gradually.

16. VFDs increase motor speed in steps. (True/False) _______________

17. Variable-frequency drive inputs can be classified as being either _______________ or _______________.

18. Which of the following devices would be considered to be digital in nature?

    a. Push button

    b. Selector switch

    c. Relay contact

    d. All of these

    Answer _______

19. A speed control potentiometer would be classified as an analog input. (True/False) _______________

20. Which of the following motor nameplate data may be programmed into a VFD?

    a. Speed

    b. Full-load current

    c. Voltage

    d. All of these

    Answer _______

21. Derating a VFD means using a smaller than normal drive in the application. (True/False) _______________

22. Derating may be required when a drive is installed at high altitude. (True/False) _______________

23. Answer the following with reference to the VFD circuit shown in Figure 10-8.

    a. The drive would be classified as a voltage source inverter type. (True/False) _______________

    b. This circuit uses _______________ for the electronic switching devices.

    c. With this arrangement, the output from the converter is a _______________ DC voltage.

    d. The inverter section takes six 60° steps to complete one 360° cycle. (True/False) _______________

    e. One disadvantage of this type of drive is cogging that may occur at _______________ speeds.

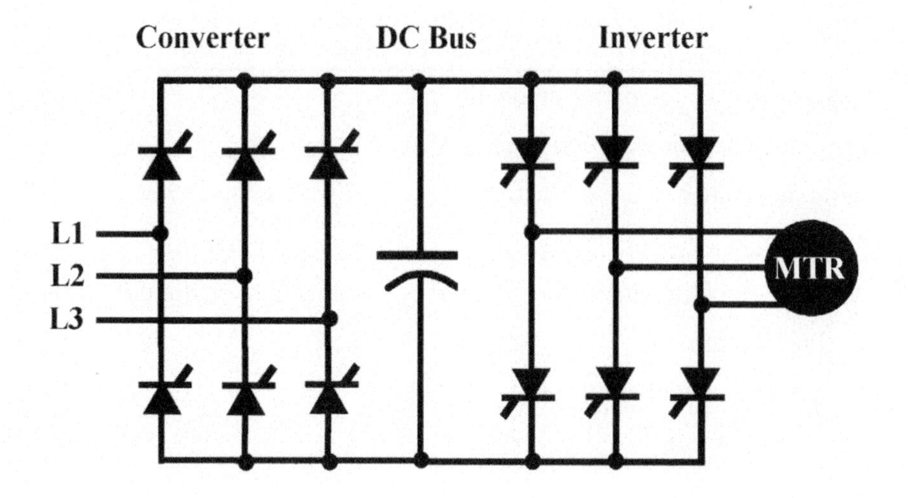

Figure 10-8 VFD circuit for Question 23.

24. The PID controller found with some VFDs automatically corrects the error between the actual motor speed and the desired setpoint speed by calculating the _______________ between the two and then outputting a corrective action.
25. Tuning the PID controller involves _______________ and _______________ adjustments that result in a fast response with minimum overshoot.
26. Which of the following would *not* be considered a VFD program parameter?
    a. Minimum and maximum speeds
    b. Acceleration and deceleration rates
    c. Type of enclosure
    d. Carrier frequency

    Answer _______
27. Which of the following types of program parameters can be changed or adjusted while the drive is running or stopped?
    a. Processor firmware
    b. Tunable on the fly
    c. Configurable
    d. Read-only

    Answer _______
28. Examples of drive display parameters include _______________, _______________, _______________, and _______________.

29. Nonconfigurable alarms can be enabled or disabled by the operator.

    (True/False) _______________

30. Analog signals are represented by two distinct amplitudes.

    (True/False) _______________

31. A 4–20 current loop is more immune to noise than a 0–10 V signal.

    (True/False) _______________

32. The sensor of a 4–20 mA current loop

    a. converts the signal into 4–20 mA value
    b. powers the transmitter
    c. measures the value of the process variable
    d. interprets and reacts to the 4–20 mA signal

33. The power source of a 4–20 mA current loop

    a. converts the signal into 4–20 mA value
    b. powers the transmitter
    c. measures the value of the process variable
    d. interprets and reacts to the 4–20 mA signal

34. The transmitter of a 4–20 mA current loop

    a. converts the signal into 4–20 mA value
    b. powers the transmitter
    c. measures the value of the process variable
    d. interprets and reacts to the 4–20 mA signal

35. The receiver of a 4–20 mA current loop

    a. converts the signal into 4–20 mA value
    b. powers the transmitter
    c. measures the value of the process variable
    d. interprets and reacts to the 4–20 mA signal

36. A 4–20 mA transmitter has an input range of 0–200 °F. If the loop current is 12 mA, the temperature indicated would be

    a. 100 °F
    b. 25 °F
    c. 150 °F
    d. 175 °F

PART 3 Quiz: DC Motor Drive Fundamentals

1. The speed of a DC motor is _______________ proportional to the armature voltage and _______________ proportional to the field current.
2. To change the direction of rotation of a DC motor, you
   a. Reverse the armature polarity
   b. Reverse the field polarity
   c. Reverse either the armature or field polarity
   d. Reverse both the armature and field polarity

   Answer _______
3. Answer the following with reference to the block diagram of the DC drive shown in Figure 10-9.
   a. This drive uses _______________ semiconductor switching.
   b. The shunt field is supplied with a variable DC voltage. (True/False) _______________
   c. The power converter block converts the supply voltage to an adjustable _______________ _______________ voltage.
   d. The current feedback signal gauges motor _______________, while the voltage feedback signal gauges motor _______________.

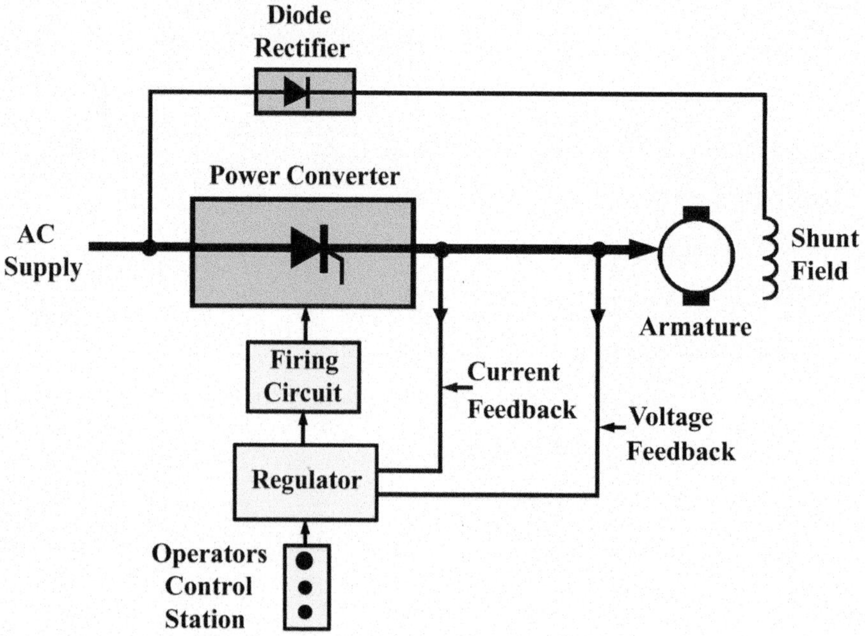

Figure 10-9 DC drive for Question 3.

4. Complete the wiring for the DC drive circuit shown in Figure 10-10.

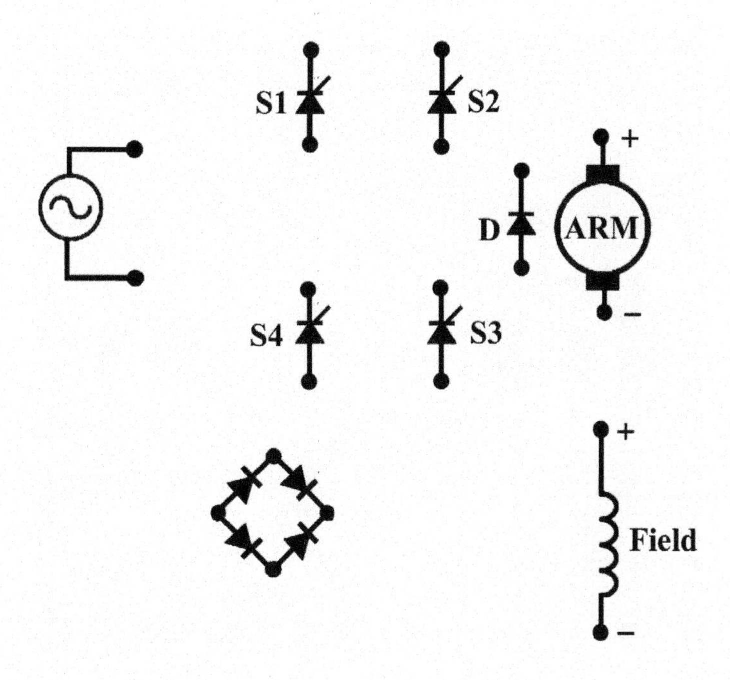

Figure 10-10 DC drive for Question 4.

5. a. Complete the wiring of the armature circuit for the nonregenerative DC drive circuit shown in Figure 10-11.

   b. In this circuit, braking is achieved by dissipating the braking energy through the DB resistor as _______________.

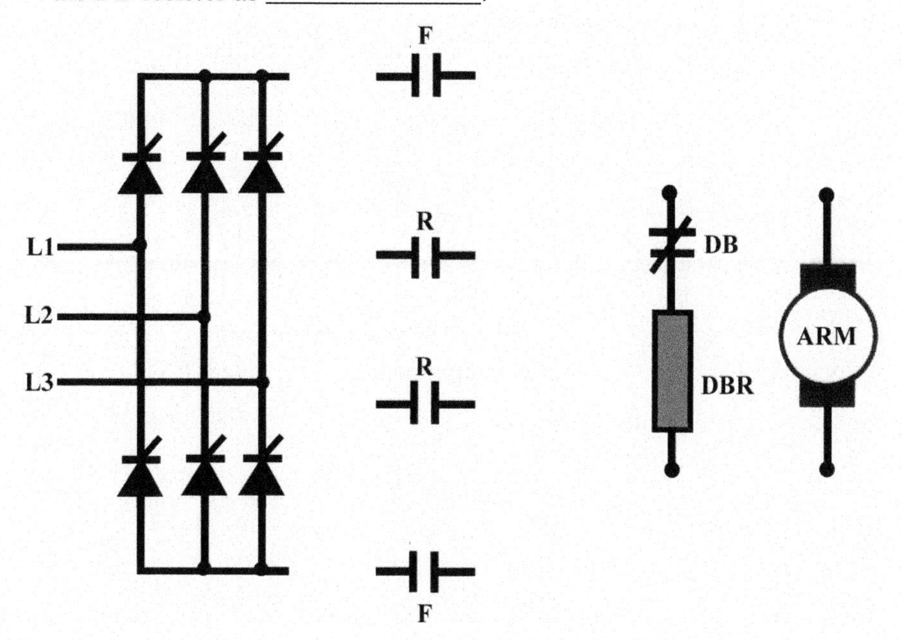

Figure 10-11 Armature circuit for Question 5.

6. a. Complete the wiring of the armature circuit for the regenerative DC drive circuit shown in Figure 10-12.
   b. In this circuit, braking is achieved by conducting the braking energy back to the _______________.

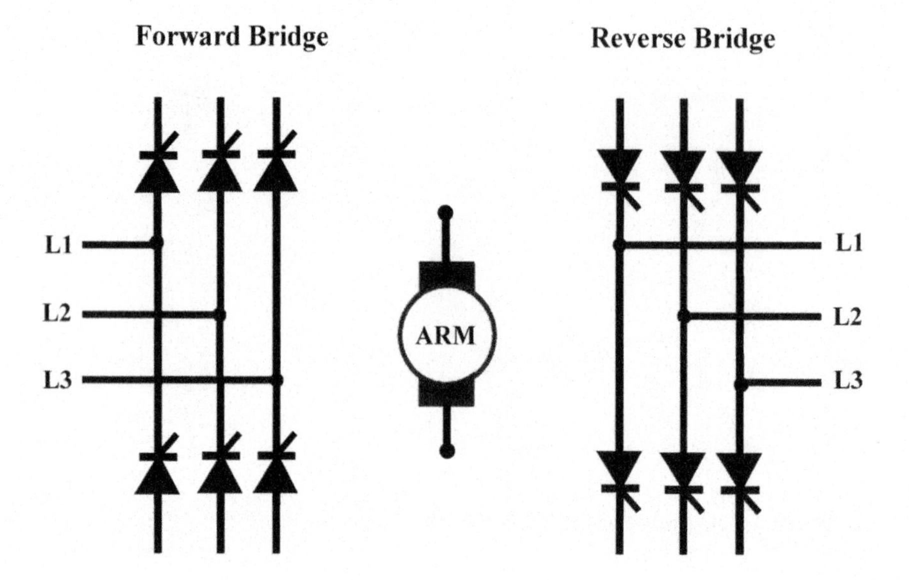

Figure 10-12 Armature circuit for Question 6.

7. An overhauling load
   a. Results in generator action within a motor
   b. Occurs whenever the inertia of the load is less than that of the motor rotor
   c. Occurs whenever the inertia of the load is more than that of the motor rotor
   d. Both a and c

   Answer _______
8. Programming parameters associated with DC drives are similar to those associated with AC drives. (True/False) _______________
9. Which of the following DC drive parameters sets the time it takes to change speeds from one setting to another?
   a. Acceleration time
   b. Deceleration time
   c. IR compensation
   d. Maximum speed

   Answer _______

PART 4 Quiz: Programmable Logic Controllers (PLCs)

1. A programmable logic controller (PLC) is an industrial-grade ______________ capable of being programmed to perform control functions.
2. PLCs are the most widely used process control technology. (True/False) ______________
3. Identify the components of the block diagram of a PLC shown in Figure 10-13.

   a. ______________ ______________ ______________

   b. ______________ ______________

   c. ______________ ______________

   d. ______________ ______________

   e. ______________ ______________

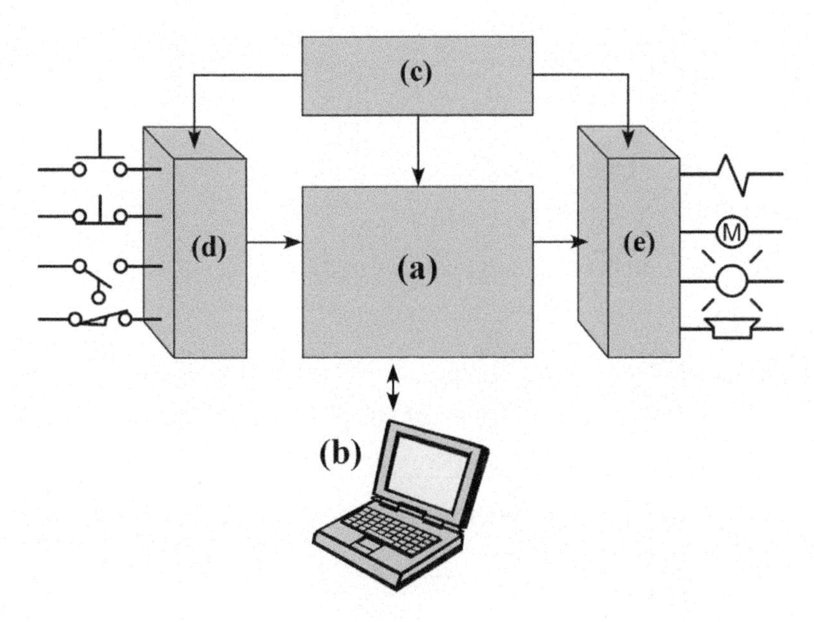

Figure 10-13 Block diagram for Question 3.

4. The programming device must be connected to the PLC at all times in order to run the program. (True/False) ______________
5. PLCs were originally designed as replacements for

   a. Microcomputers

   b. Relay control panels

   c. Analog controllers

   d. Digital controllers

   Answer _______

6. The central processing unit or processor of a PLC
   a. Looks at the inputs, makes the decision provided by the program, and sets the status of the outputs
   b. Looks at the outputs, makes the decision provided by the program, and sets the status of the inputs
   c. Serves only to store the program in memory
   d. Understands only ladder logic

   Answer _______
7. The power supply required to operate the PLC processor and I/O modules units is
   a. Low-voltage AC
   b. High-voltage AC
   c. Low-voltage DC
   d. High-voltage DC

   Answer _______
8. PLC input modules
   a. Receive signals from field devices
   b. Convert input signals to logic signals
   c. Provide isolation between field devices and PLC circuitry
   d. All of these

   Answer _______
9. PLC output modules convert signals from the CPU that are used to control output field devices. (True/False) _______________
10. Which of the following is *not* a function of the PLC programming device?
    a. To enter the user program
    b. To change the user program
    c. To run the user program
    d. To monitor the user program

    Answer _______
11. A nonmodular or fixed PLC consists of a _______________, _______________, and _______________ in a single unit.
12. The chassis of a modular PLC is divided by compartments into which the separate _______________ can be plugged.
13. A _______________ is a user-developed series of instructions that direct the PLC to execute actions.

14. The relay ladder logic (RLL) programming language is based on electromagnetic relay control. (True/False) _______________
15. A relay ladder logic program graphically represents rungs of _______________, _______________, and _______________, _______________ blocks.
16. Answer the following with reference to the PLC motor start/stop program shown in Figure 10-14.
    a. What is the address associated with each of the following field devices?
       (1) Stop button _______________ (2) Start button _______________
       (3) Starter coil _______________ (4) Starter coil contact _______________
    b. What is the instruction associated with each of the following field devices?
       (1) Stop button _______________ (2) Start button _______________
       (3) Starter coil _______________ (4) Starter coil contact _______________
    c. What is the state (true/false) of each of the following when the motor starter coil is deenergized?
       (1) I1 _______________ (2) I2 _______________ (3) I3 _______________
    d. What is the state (true/false) of each of the following when the motor starter coil is energized?
       (1) I1 _______________ (2) I2 _______________ (3) I3 _______________

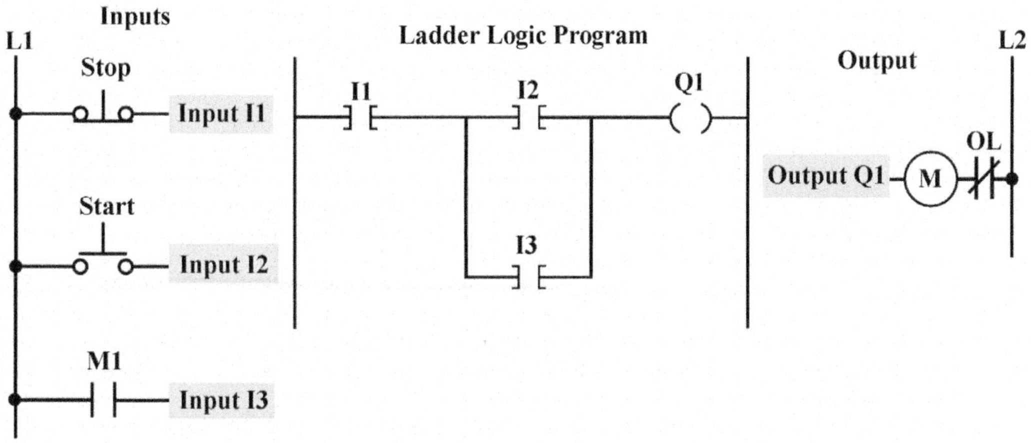

Figure 10-14 Start/stop program for Question 16.

17. The addressing format used for PLCs is the same for all manufacturers. (True/False) _______________
18. In RLL programs outputs are represented by _______________ symbols.

19. The term _______________-wired refers to logic control functions that are determined by the way the devices are interconnected.
20. From a safety standpoint, only normally _______________ push buttons should be utilized for motor stop buttons.
21. Complete the wiring for the PLC shown in Figure 10-15 programmed for a motor start/stop circuit similar to that shown in the text, using the following addresses:

    Normally open start button—I2

    Normally closed stop button—I6

    Starter coil—Q4

    Normally open starter contact (M1)—I8

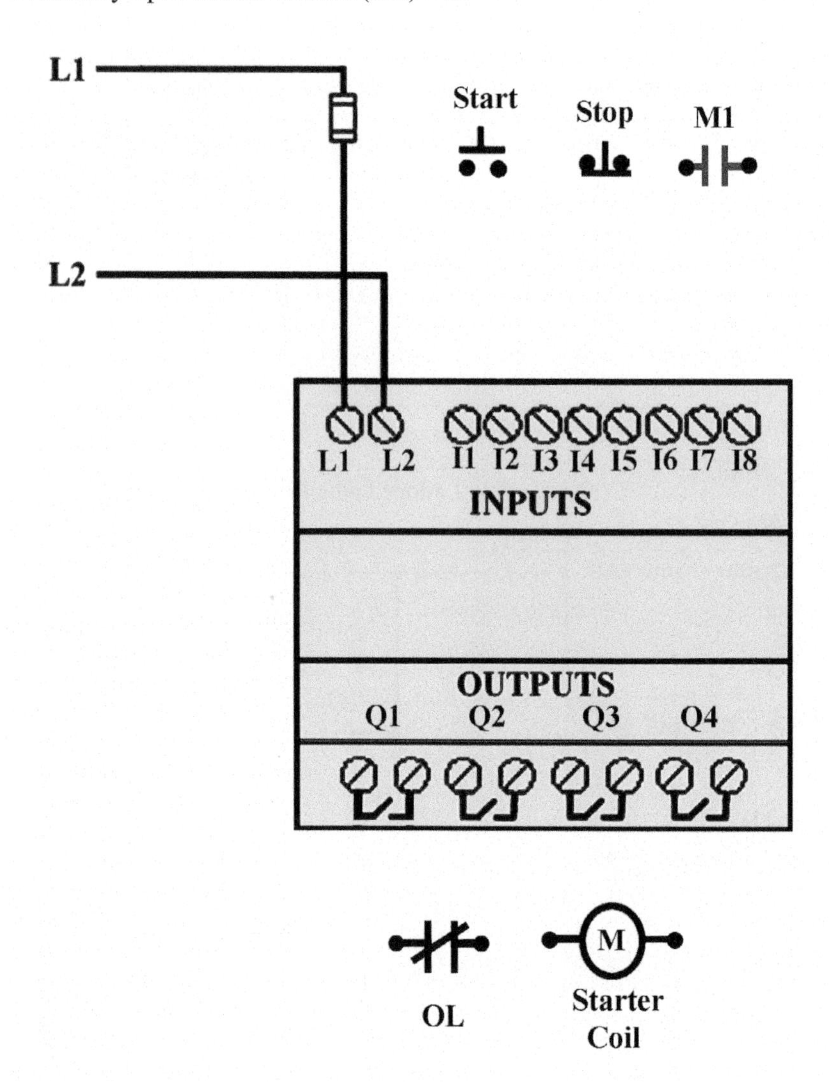

Figure 10-15 PLC diagram for Question 21.

22. One additional start and stop button is to be added to the original PLC motor circuit containing remote standby and run pilot lights outlined in the text and shown in Figure 10-16. Complete the ladder logic program for these changes.

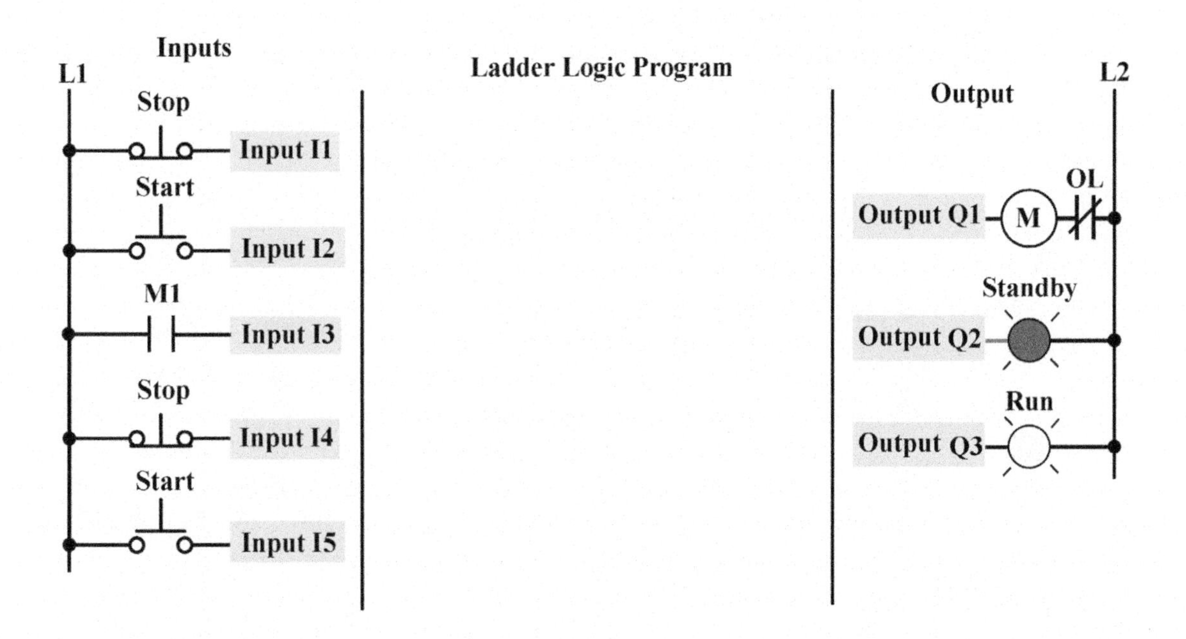

Figure 10-16 PLC motor circuit for Question 22.

23. PLC timers
    a. Are input instructions
    b. Are output instructions
    c. Provide the same function as electromechanical timing relays
    d. Both b and c

    Answer _______

24. Which of the following PLC timers is used to delay the start of a machine process?
    a. Retentive timer
    b. On-delay timer
    c. Off-delay timer
    d. On/off timer

    Answer _______

25. Which of the following changes state whenever the accumulated value of the PLC timer reaches the preset value?
    a. Enable bit
    b. Done bit
    c. Time base
    d. Timer number

    Answer _______

26. Answer the following with reference to the diagram of the PLC timer program shown in Figure 10-17.
    a. This timer would be classified as an ______________-delay type.
    b. The time-delay period for this timer is set for ______________ seconds.
    c. Pilot light output B will be on at all times that the switch is opened. (True/False) ______________
    d. Pilot light output C will come on when the switch is closed and the time-delay period has elapsed. (True/False) ______________
    e. The timer's accumulated value resets to zero any time the switch is opened. (True/False) ______________

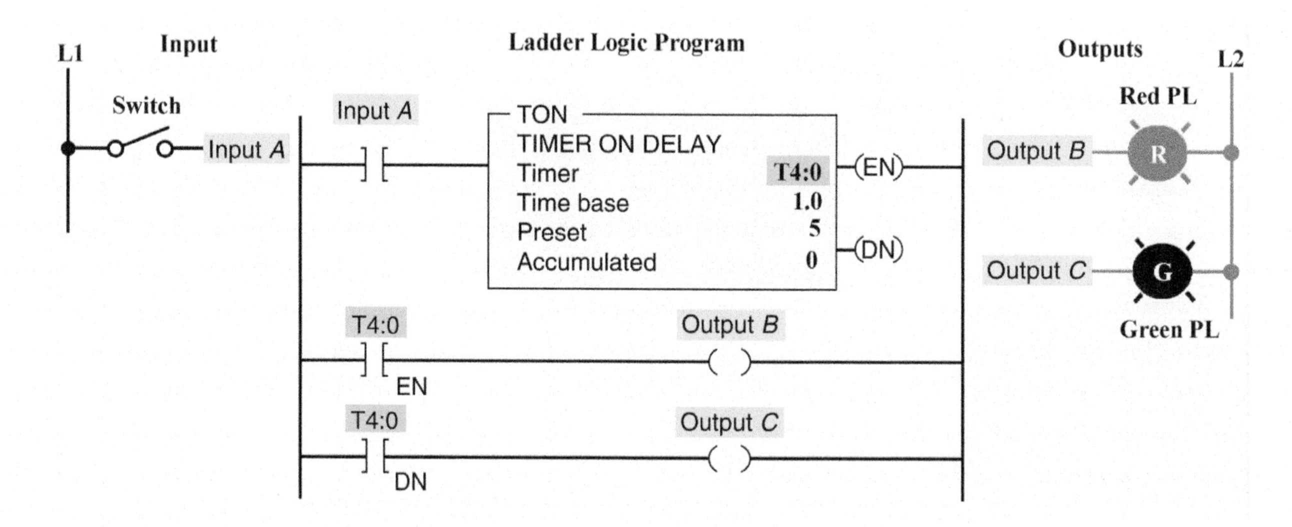

Figure 10-17 PLC timer program for Question 26.

27. The correct address for a pilot light connected to terminal 5, slot 3, of a single-rack Allen-Bradley SCL 500 controller would be

    a. O:3/5
    b. O:5/3
    c. I:3/5
    d. I:5/3

    Answer _______

28. Programmed counters can

    a. Count up
    b. Count down
    c. Be combined to count up and down
    d. All of these

    Answer _______

29. Answer the following with reference to the diagram of the PLC timer counter program shown in Figure 10-18.

    a. This counter would be classified as an ______________ counter.
    b. The counter is set to actuate the done bit instruction whenever a count of ______________ is reached.
    c. The on-to-off transition pulses of push button ______________ are counted by the counter.
    d. What would be the state of the pilot lights (on or off?) after the following sequence?
        - PB2 momentarily actuated
        - PB1 actuated on to off eight times
        - PB2 momentarily actuated
        - PB1 actuated on to off two times

        (1) Red pilot light ______________

        (2) Green pilot light ______________

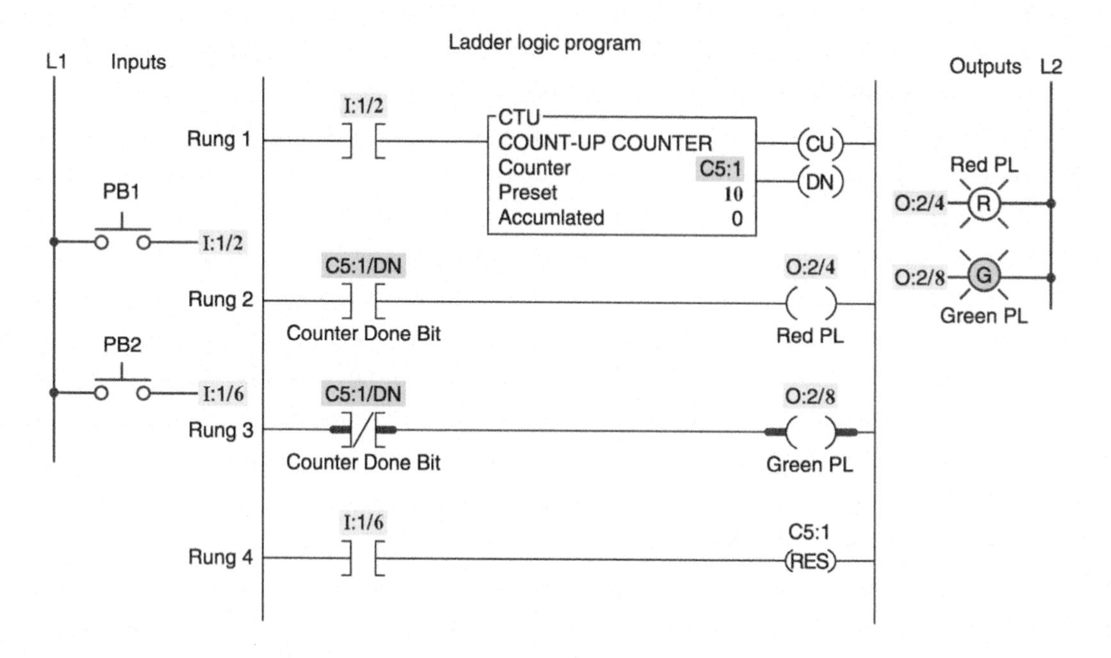

Figure 10-18 PLC timer counter program for Question 29.

30. Complete the wiring for the PLC shown in Figure 10-19 using the addressing and program of Question 29.

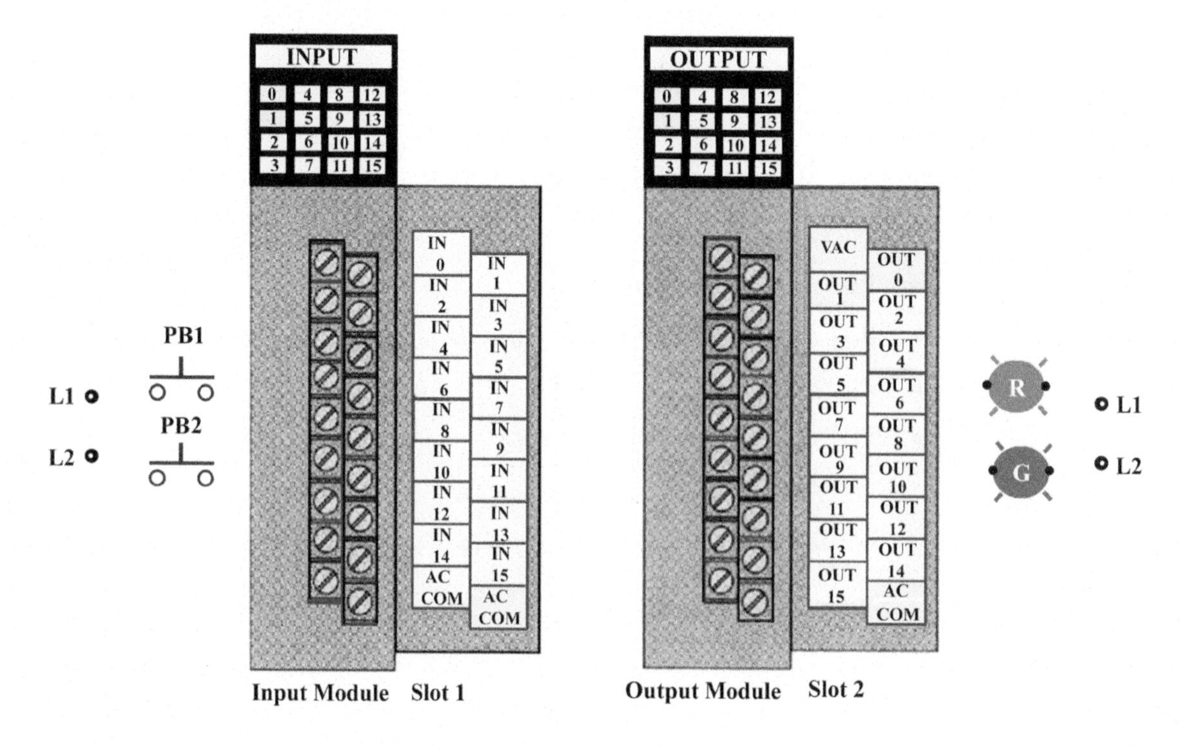

Figure 10-19 PLC diagram for Question 30.

31. A watchdog timer is used to monitor the scan process of a PLC.

(True/False) _______________

32. For an input module, where is the problem most likely to reside, if the status indicator and ladder instructions agree, but not with the sensor condition?

a. Field device or wiring
b. Input module
c. Processor
d. PLC program

33. For an input module, where is the problem most likely to reside, if the status indicator and ladder instructions agree, but not with the sensor condition?

a. Field device or wiring
b. Input module
c. Output module
d. PLC program

34. For an output module, where is the problem most likely to reside, if the field device condition and status indicator agree, but the output instruction does not?

a. Field device or wiring
b. Input module
c. Output module
d. PLC program

Part 2

“THE CONSTRUCTOR” SIMULATION LAB MANUAL

2 Understanding Electrical Diagrams

CIRCUIT ANALYSIS ASSIGNMENTS

To download the latest version of The Constructor, go to the registration website on the access card provided with the Activities Manual.

1. The purpose of this assignment is to analyze the operation of the **two-wire motor control circuit** shown. Download the analysis assignment circuit file **CH2 Assign 01** from the **Diagrams-Analysis-NEMA** folder in The Constructor software folder. Click on the Main Power Switch (ON/OFF) located on the left side of the toolbar. Left-click on the circuit breaker to close it. Left-click on the temperature switch to alternate between closed and open states.

   a. Outline what happens when the temperature switch is closed.

   ____

   ____

   b. Outline what happens when the temperature switch is opened.

   ____

   c. With the temperature switch closed, open and close the circuit breaker to remove and apply power to the circuit. Outline the reaction of the motor when the power is removed and reapplied.

   ____

   ____

   d. With the motor operating, press shift and left-click on the motor to simulate a motor overload condition. Outline how the control circuit reacts to this overload condition.

   ____

   ____

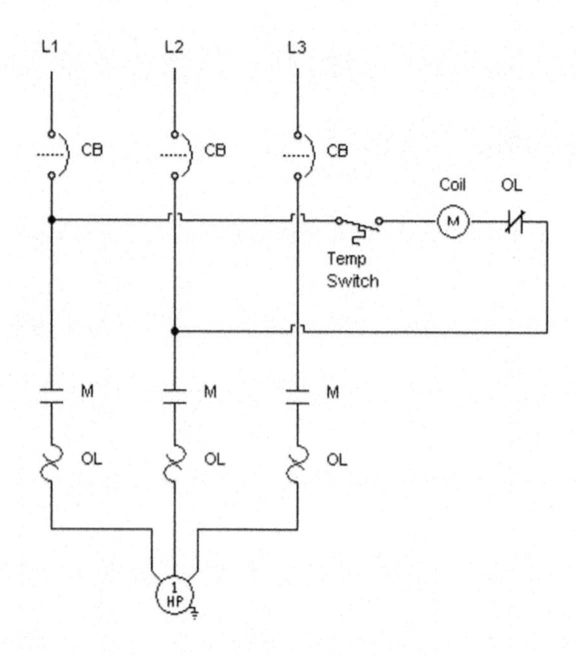

Two-wire control circuit.

2. The purpose of this assignment is to analyze the operation of the **three-wire motor control circuit** shown. Download the analysis assignment circuit file **CH2 Assign 02** from the **Diagrams-Analysis-NEMA** folder. Click on the Main Power Switch (ON/OFF) and left-click on the main switch to close it. Left-click on the start and stop push buttons to operate them.

   a. Outline what happens when the start button is momentarily closed.

   ____

   ____

   b. With the motor operating, outline what happens when the stop button is momentarily opened.

   ____

   ____

   c. With the motor operating, open and close the main switch to remove and apply power to the circuit. Outline the reaction of the motor when the power is removed and reapplied.

   ____

   ____

d. With the motor operating, press shift and left-click on the motor to simulate a motor overload condition. Outline how the motor circuit reacts to this overload condition.

____

____

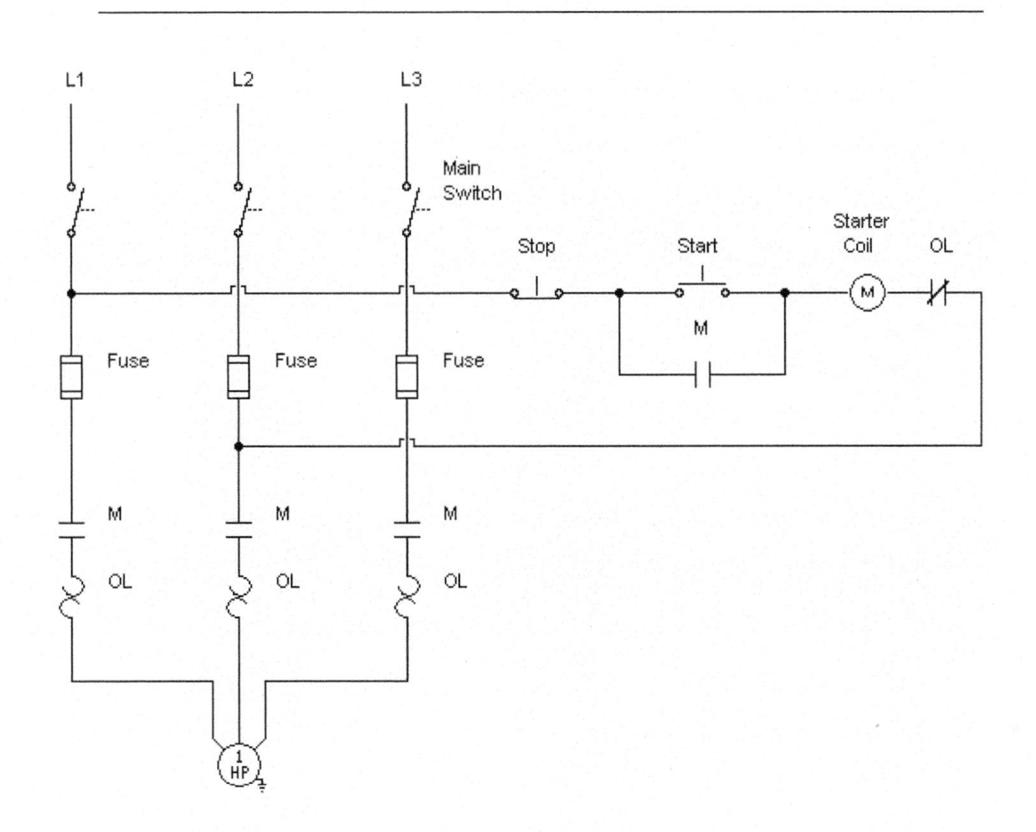

Three-wire control circuit.

"THE CONSTRUCTOR" CIRCUIT TROUBLESHOOTING ASSIGNMENTS

1. This assignment involves the **troubleshooting** of a two-wire **inoperative** motor control circuit. Download the troubleshooting assignment circuit file **CH2 T01** from the **Diagrams-Troubleshooting-NEMA** folder.
   a. Turn on the Main Power Switch. Use the test probe Power function to record the (ON/OFF) power state of the following test points:
      i. Line side of the CB ____________________
      ii. Load side of the circuit breaker (breaker open) ____________________
      iii. Load side of the circuit breaker (breaker closed) ____________________
      iv. Power across M coil (Temp Switch open) ____________________
      v. Power across M coil (Temp Switch closed) ____________________
      vi. Power at motor terminals (Temp Switch open) ____________________
      vii. Power at motor terminals (Temp Switch closed) ____________________
   b. Based on your readings, what is the most likely problem?

      ____
   c. Explain what specific measurements lead you to conclude this?

      ____

      ____

      (Turn the Main Power Switch off, replace the defective component, and turn the Main power back on and operate the circuit to verify your answer.)

2. This assignment involves the **troubleshooting** of a three-wire **faulty operative** motor control circuit. Download the troubleshooting assignment circuit file **CH2 T02** from the **Diagrams-Troubleshooting-NEMA** folder.
   a. Close the main switch and summarize the faulty operating condition that exists.

      ____

      ____

   b. Open the Main switch. Left-click in the wire terminal connection between the Start pushbutton and the M auxiliary contact in order to open the circuit between the two. Use the test probe Continuity function to record the (Yes/NO) continuity state of the following test points:
      i. Across each of the main M motor contacts. ____________
      ii. Across the auxiliary M contact ____________
      iii. Across the Stop pushbutton in the deactivated position. ____________
      iv. Across the Stop pushbutton in the activated position. ____________
      v. Across the Start pushbutton in the deactivated position. ____________
      vi. Across the Start pushbutton in the activated position. ____________
      vii. Across the OL contact. ____________

   c. Based on your readings, what is the most likely fault?

      ____

      ____

   d. What specific continuity measurement lead you to conclude this?

      ____

      ____

   e. Assume that opening the wire terminal connection between the Start pushbutton and the M auxiliary contact was not done before taking your continuity measurements. What false continuity reading would be evident?

      ____

      ____

      ____

      (Turn the Main Power Switch off and replace the defective component. Turn the Main power back, left-click terminal wire to re-establish the connection, and operate the circuit to verify your answer.)

3 Motor Transformers and Distribution Systems

CIRCUIT ANALYSIS ASSIGNMENTS

1. The purpose of this assignment is to analyze the operation of the **properly grounded control transformer circuit** shown. Download the analysis assignment circuit file **CH3 Assign 01** from the **Diagrams-Analysis-NEMA** folder. Click on the Main Power Switch (ON/OFF) located on the left side of the toolbar. Left-click on the circuit breaker to close it and observe the normal operation of the circuit.

   a. How is the motor normally started?

   ____

   b. How is the motor normally stopped?

   ____

   c. According to the control transformer primary connection, would the primary line voltage be 480 or 240 V? Why?

   ____

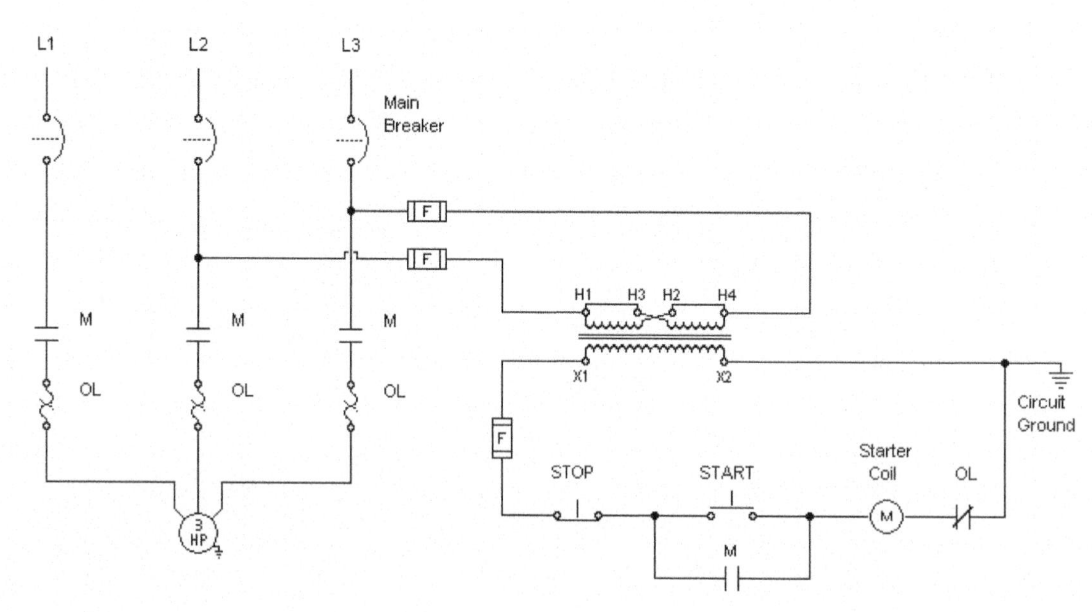

Control transformer circuit.

2. The purpose of this assignment is to analyze the operation of the **properly grounded control transformer circuit that experiences a ground fault condition** shown. Download the analysis assignment circuit file **CH3 Assign 02** from the **Diagrams-Analysis-NEMA** folder. Click on the Main Power Switch located on the left side of the toolbar. Left-click on the circuit breaker to close it and observe the operation of the circuit with a ground fault at the connection between the start and stop push buttons. What happens as soon as the main circuit breaker is closed? Why?

____

____

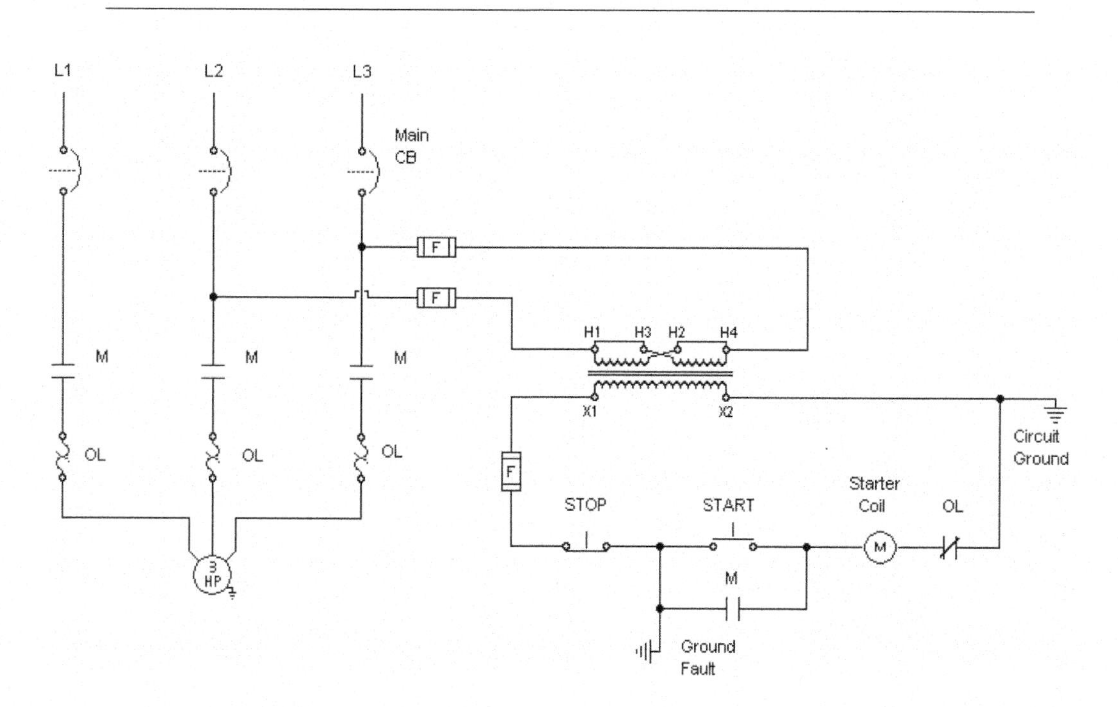

Ground-faulted control transformer circuit.

3. The purpose of this assignment is to analyze the operation of the ***incorrectly grounded*** **control transformer circuit that experiences a ground fault condition** shown. Download the analysis assignment circuit file **CH3 Assign 03** from the **Diagrams-Analysis-NEMA** folder. Click on the Main Power Switch located on the left side of the toolbar. Left-click on the circuit breaker to close it and observe the operation of the circuit with the same ground fault and the transformer incorrectly grounded at X1 instead of X2.

   a. What happens when the main circuit breaker is closed and the start button is momentarily clicked closed?

   ____

   ____

   b. What happens when the motor is running and the stop button is clicked open? Why?

   ____

   ____

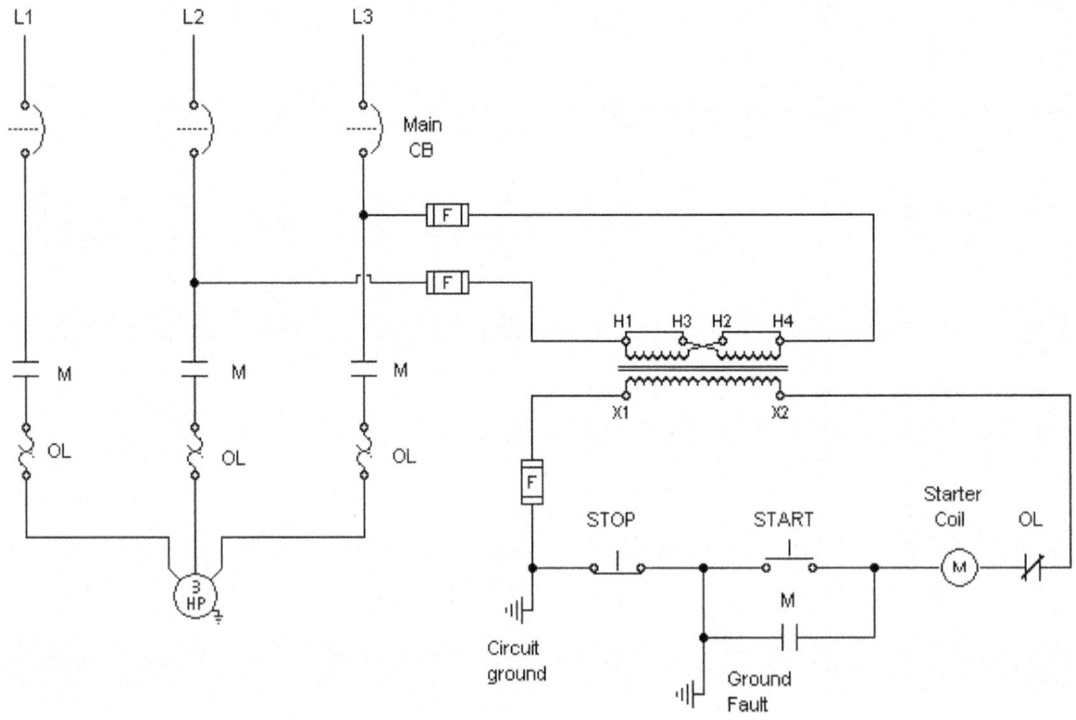

Incorrectly connected transformer ground circuit.

TROUBLESHOOTING ASSIGNMENTS

1. This assignment involves the **troubleshooting** of an **inoperative** motor circuit equipped with a control transformer. Download the troubleshooting assignment circuit file **CH3 T01** from the **Diagrams-Troubleshooting-NEMA** folder.
   a. Turn on the Main Power Switch. Close the main breaker and with the test probe set to power record the (ON/OFF) power state of the following test points:
      i. Across the three leads on the load side of the circuit breaker. _________
      ii. H1 to H4 of the transformer primary. _________
      iii. X1 to X2 of the transformer secondary. _________
      iv. X1 to both sides of the OL contact. _________
      v. X2 to the line side of the Stop button. _________
   b. Based on your readings, what is the most likely problem?

      ____

   c. Explain what specific measurements lead you to conclude this?

      ____

      (Turn the Main Power Switch off, replace the defective component, and turn the Main power back on and operate the circuit to verify your answer.)

2. This assignment involves the **troubleshooting** of an **inoperative** motor circuit equipped with a control transformer. Download the troubleshooting assignment circuit file **CH3 T02** from the **Diagrams-Troubleshooting-NEMA** folder.

   a. Turn on the Main Power Switch. Close the main breaker and with the test probe set to power record the (ON/OFF) power state of the following test points:

      i. Across the three leads on the load side of the circuit breaker. _________

      ii. H1 to H4 of the transformer primary. _________________

      iii. X1 to X2 of the transformer secondary. _________________

      iv. X1 to both sides of the OL contact. ____________________

      v. X1 to both sides of coil M. ____________________

   b. Open the main circuit breaker and with the test probe set to continuity record the (YES/NO) continuity state of the following test points:

      i. Across the OL contact _______________

      ii. Across the start pushbutton (button not activated). ________________

      iii. Across the start pushbutton (button activated). ________________

      iv. Across the stop pushbutton (button not activated). ______________

      v. Across the stop pushbutton (button activated). ________________

      vi. Across fuse F3. _________________

   c. Based on your readings, what is the most likely problem?

      ____

   d. Explain what specific measurements lead you to conclude this?

      ____

      (Turn the Main Power Switch off, replace the defective component, and turn the Main power back on and operate the circuit to verify your answer.)

4 Motor Control Devices

CIRCUIT ANALYSIS ASSIGNMENTS

1. The purpose of this assignment is to analyze the operation of the **break-make pushbutton circuit** shown. Download the analysis assignment circuit file **CH4 Assign 01** from the **Diagrams-Analysis-NEMA** folder. Click on the Main Power Switch (ON/OFF) located on the left side of the toolbar. Left-click on the circuit breaker to close it. Operate the break-make pushbutton by left-clicking on it.
   a. What is the state of the pilot lights when the button is not actuated?

   ____

   b. What is the state of the pilot lights when the button is actuated?

   ____

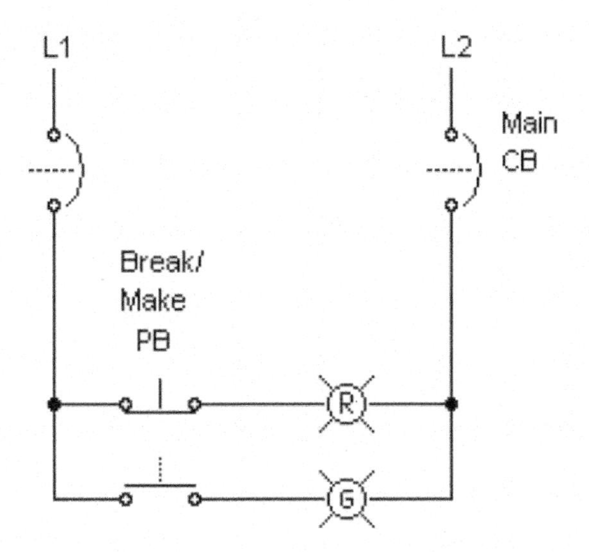

Break-make pushbutton circuit.

2. The purpose of this assignment is to analyze the operation of the **emergency stop pushbutton circuit** shown. Download the analysis assignment circuit file **CH4 Assign 02** from the **Diagrams-Analysis-NEMA** folder. Click on the Main Power Switch (ON/OFF) located on the left side of the toolbar. Left-click on the circuit breaker to close it. Operate the circuit control pushbuttons and observe what happens.

   a. What type of contacts (maintained or momentary) are used for the start and stop pushbuttons?

   ____

   b. With the motor running, actuate the E-Stop button. How is the motor restarted once the E-Stop button is pressed?

   ____

   ____

   c. What type of contact (maintained or momentary) is used for the emergency stop pushbutton?

   ____

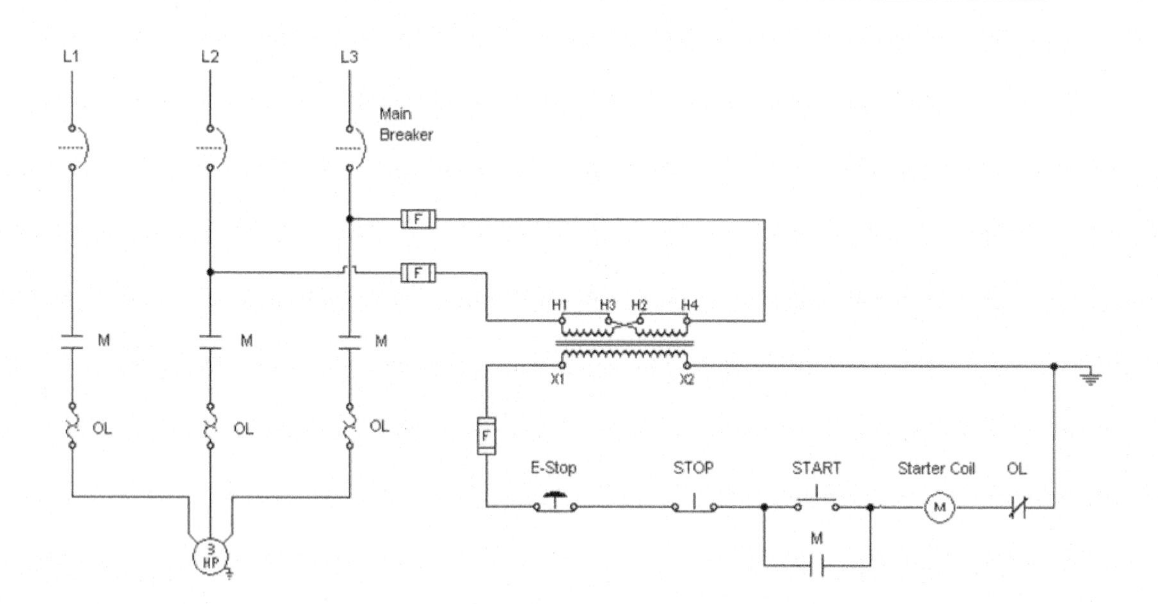

Emergency stop pushbutton circuit.

3. The purpose of this assignment is to analyze the operation of **pilot lights used in conjunction with a magnetic motor starter circuit** shown. Download the analysis assignment circuit file **CH4 Assign 03** from the **Diagrams-Analysis-NEMA** folder. Click on the Main Power Switch (ON/OFF) located on the left side of the toolbar. Left-click on the main breaker to close it.

   a. What is the initial status of the pilot lights?

   ____

   b. Start the motor. What is the new status of the pilot lights?

   ____

   c. Right-click on the NC auxiliary M contact and lock it in the open position. What effect does this has on the operation of the pilot lights?

   ____

   ____

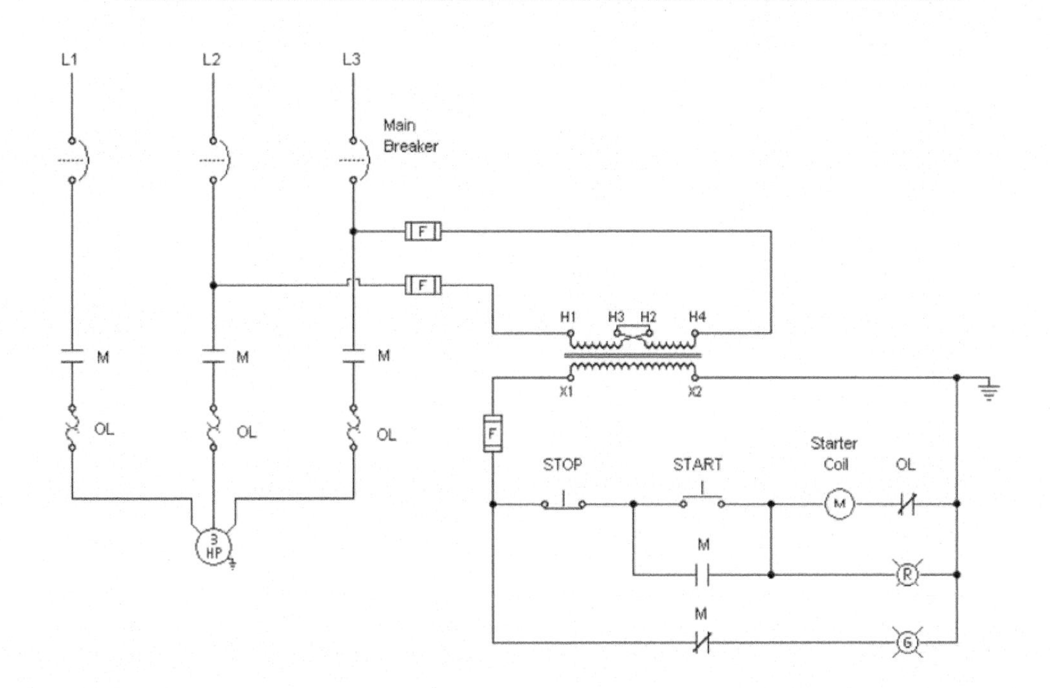

Motor starter pilot light circuit.

4. The purpose of this assignment is to analyze the operation of the **push-to-test pilot light circuit** shown. Download the analysis assignment circuit file **CH4 Assign 04** from the **Diagrams-Analysis-NEMA** folder. Click on the Main Power Switch (ON/OFF) located on the left side of the toolbar. Left-click on the main breaker to close it.

   a. What is the pilot light connected to indicate?

   ____

   b. Start the motor. What is the status of the pilot light?

   ____

   c. With the motor running, left-click the pilot light. What is the status of the pilot light?

   ____

   d. Stop the motor. What is the status of the pilot light?

   ____

   e. With the motor stopped, left-click the pilot light. What is the status of the pilot light?

   ____

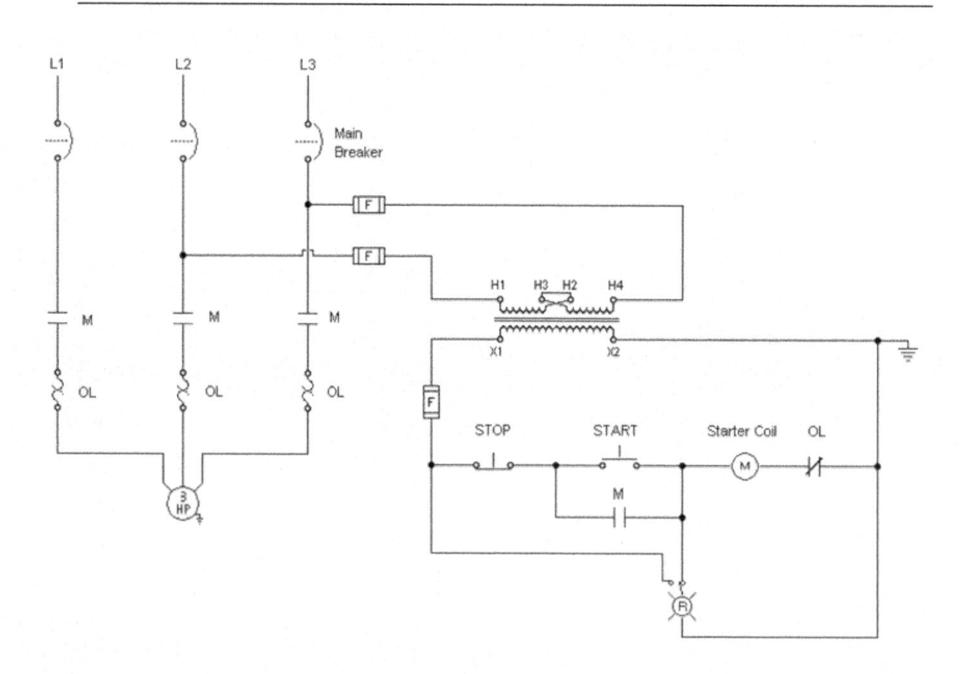

Push-to-test pilot light circuit.

5. The purpose of this assignment is to analyze the operation of the **three-position selector switch circuit** shown. Download the analysis assignment circuit file **CH4 Assign 05** from the **Diagrams-Analysis-NEMA** folder. Click on the Main Power Switch (ON/OFF) located on the left side of the toolbar. Left-click on the main breaker to close it.
   a. With the selector switch in the manual (M) position, what device controls the operation of the motor?

   ___

   b. Right-click the selector switch to the OFF position. Why will the motor not operate with the switch in this position?

   ___

   c. Right-click the selector switch to the auto (A) position. What device controls the operation of the motor with the switch in this position?

   ___

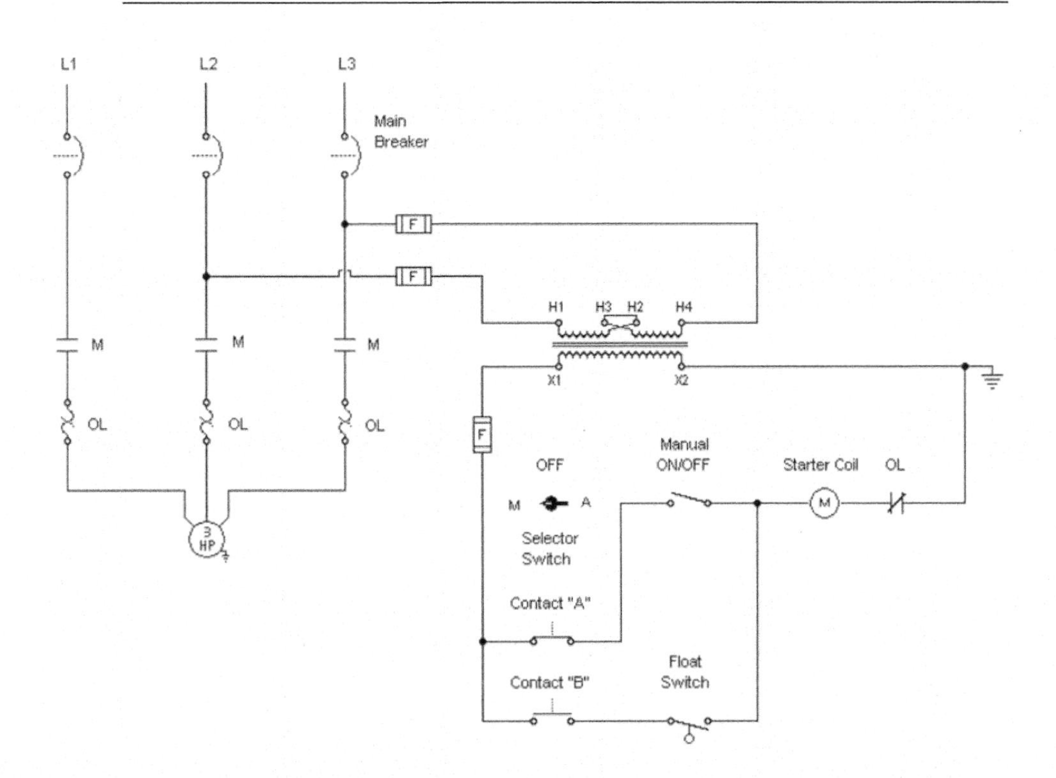

Three-position selector switch circuit.

6. The purpose of this assignment is to analyze the operation of the **two-pole limit switch circuit** shown. Download the analysis assignment circuit file **CH4 Assign 06** from the **Diagrams-Analysis-NEMA** folder. Click on the Main Power Switch (ON/OFF) located on the left side of the toolbar. Left-click on the main breaker to close it.

   a. What is the initial state of the pilot lights?

   ____

   b. Left-click the limit switch icon to activate the contacts. What is the state of the pilot lights when the limit switch is actuated?

   ____

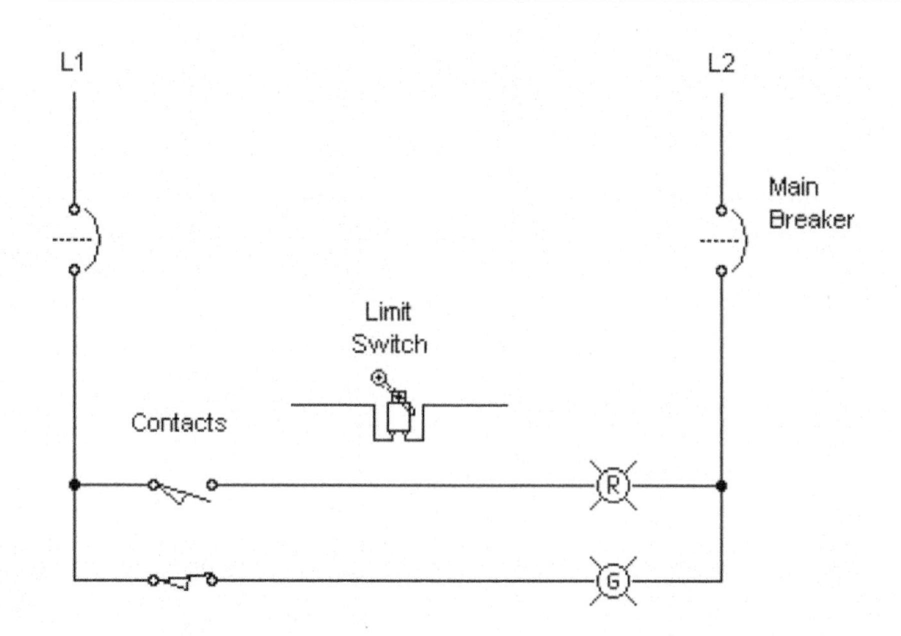

Two-pole limit switch circuit.

7. The purpose of this assignment is to analyze the operation of the **limit switch circuit used to provide overtravel protection** shown. Download the analysis assignment circuit file **CH4 Assign 07** from the **Diagrams-Analysis-NEMA** folder. Click on the Main Power Switch (ON/OFF) located on the left side of the toolbar. Left-click on the main breaker to close it.

   a. Left-click on the forward button to operate the motor in the forward direction. While the motor is operating in the forward direction, left-click on the reverse button. What prevents the reverse coil from energizing?

   ____

   b. While the motor is operating in the forward direction, actuate the forward overtravel limit switch. Why can only the reverse coil now be energized?

   ____

   ____

   c. Assume the reverse overtravel limit switch is actuated and fails to return to its normal state once travel in the forward direction has been initiated. How would this affect the operation of the circuit?

   ____

   ____

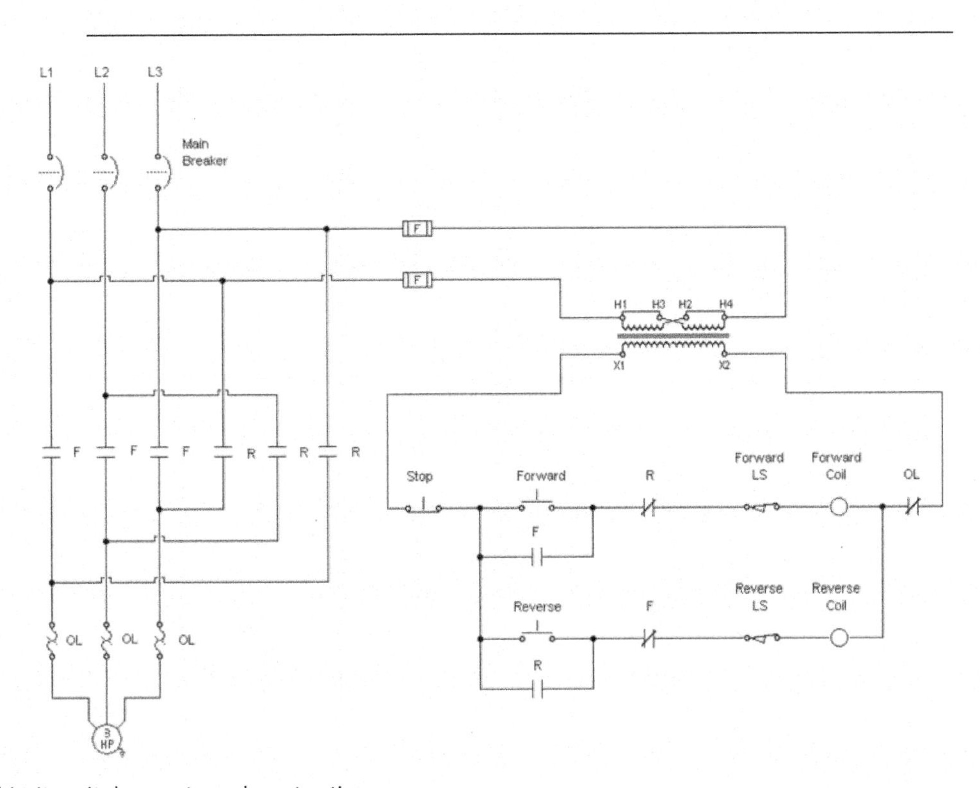

Limit switch overtravel protection.

8. The purpose of this assignment is to analyze the operation of the **temperature switch control of a motor circuit** shown. Download the analysis assignment circuit file **CH4 Assign 08** from the **Diagrams-Analysis-NEMA** folder. Click on the Main Power Switch (ON/OFF) located on the left side of the toolbar. Left-click on the manual starter to close it.
   a. Left-click the selector switch to the hand (H) position. Which selector switch contacts are closed in this position?

   ____

   b. Right-click the selector switch to the auto (A) position. Which selector switch contacts are closed in this position?

   ____

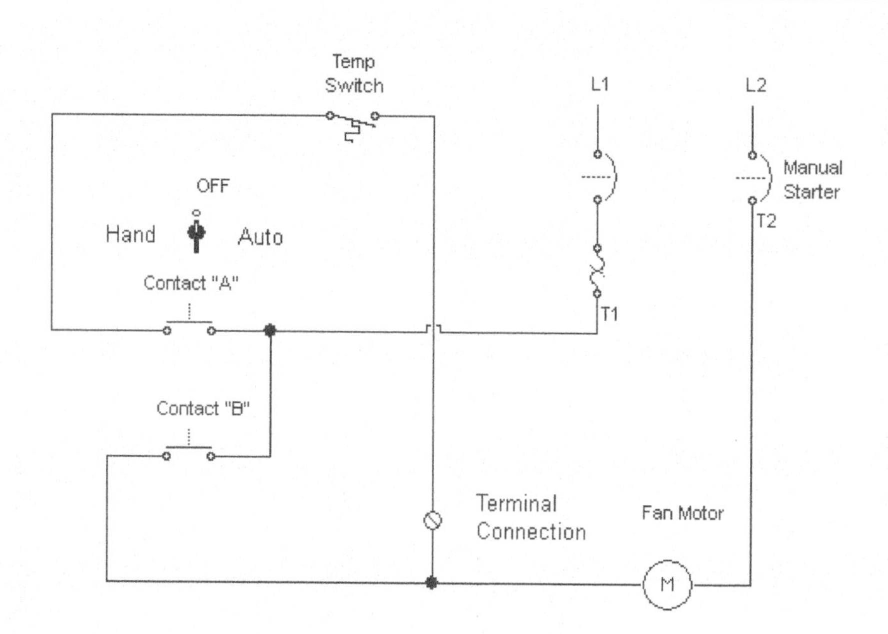

Temperature switch control of a motor circuit.

9. The purpose of this assignment is to analyze the operation of the **interposing relay in a motor control circuit** shown. Download the analysis assignment circuit file **CH4 Assign 09** from the **Diagrams-Analysis-NEMA** folder. Click on the Main Power Switch (ON/OFF) located on the left side of the toolbar. Left-click on the main breaker and on/off switch to close them.

   a. Left-click on the proximity switch to close it and outline the sequence of events that occur in order to start the motor.

   ____

   ____

   b. What is the operating voltage of the proximity switch? ____________

   c. What is the operating voltage of coil CR? ____________

   d. What is the operating voltage of contact CR? ____________

   e. What is the operating voltage of coil M? ____________

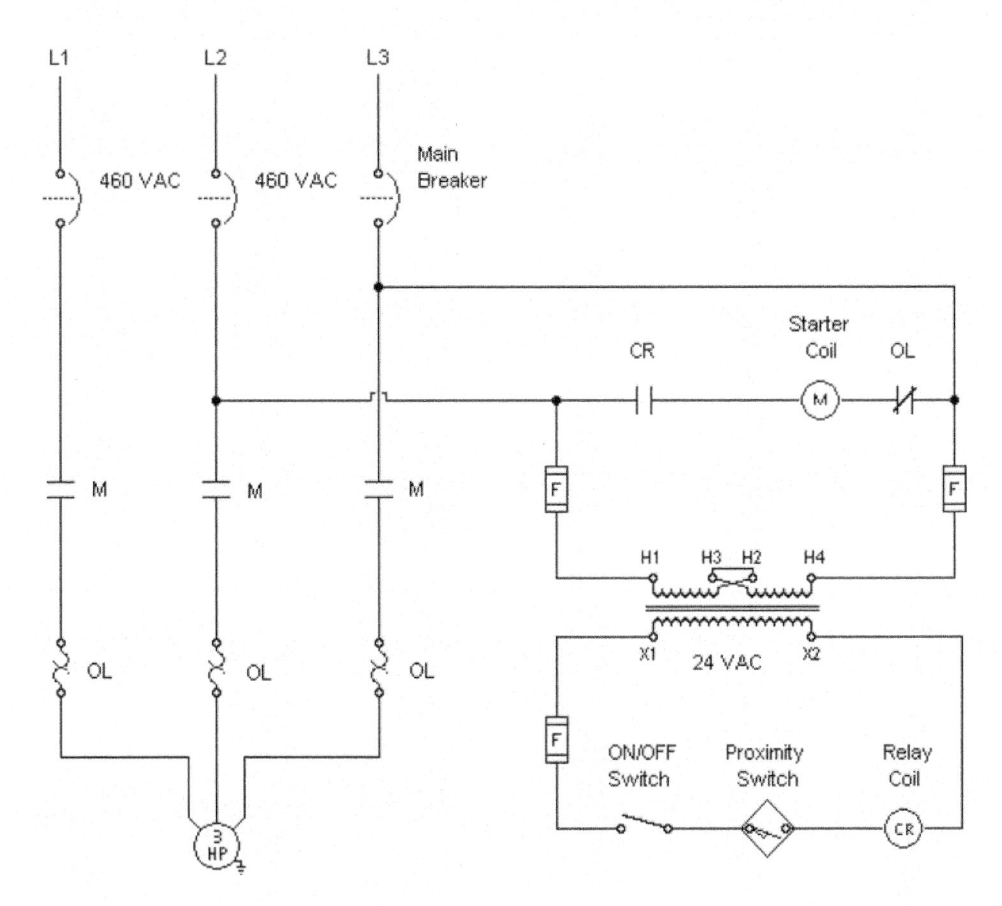

Interposing relay motor control circuit.

10. The purpose of this assignment is to analyze the operation of the **solenoid valves in a tank filling and emptying circuit** shown. Download the analysis assignment circuit file **CH4 Assign 10** from the **Diagrams-Analysis-NEMA** folder. Click on the Main Power Switch (ON/OFF) located on the left side of the toolbar.

    a. Left-click the fill button to initiate filling of the tank. Left-click on the full tank sensor to simulate a full tank. List the sequence of operation that takes place to automatically fill the tank.

    ____

    ____

    ____

    b. Once the tank is full, left-click the empty button and list the sequence of operations that takes place to automatically empty the tank.

    ____

    ____

    ____

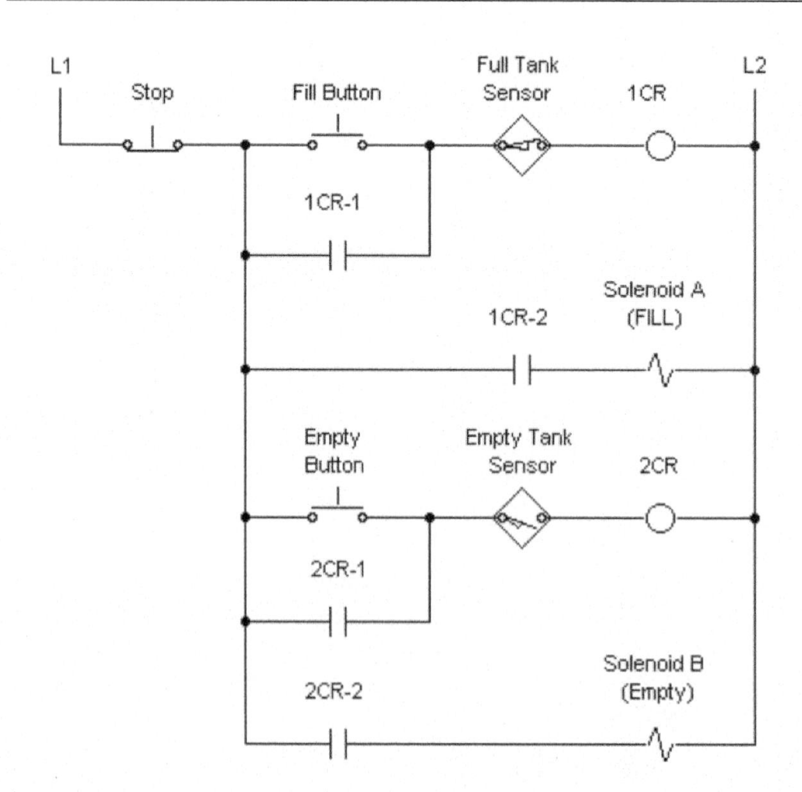

Operation of solenoid valves in a tank filling and emptying circuit.

11. The purpose of this assignment is to analyze the operation of the **alternating pumping operation control circuit** shown. Download the analysis assignment circuit file **CH4 Assign 11** from the **Diagrams-Analysis-NEMA** folder. Click on the Main Power Switch (ON/OFF).

   a. Left-click the float switch ON to initiate the pumping operation. Record the state (ON/OFF or Energized/Deenergized) of each of the following:

      i. Coil-M1 ________ iii. Coil-M2 ________ v. Coil-CR _______

      ii. Run-M1 light ______ iv. Run-M2 light ______

   b. Left-click the float switch OFF and record the state of each of the following:

      i. Coil-M1 ________ iii. Coil-M2 ________ v. Coil-CR _______

      ii. Run-M1 light ______ iv. Run-M2 light ______

   c. Left-click the float switch ON and record the state of each of the following:

      i. Coil-M1 ________ iii. Coil-M2 ________ v. Coil-CR _______

      ii. Run-M1 light ______ iv. Run-M2 light ______

   d. How is the operation of the two motors effected by each ON-OFF-ON sequencing of the float switch?

      ____

   e. What is the function of the two pilot lights?

      ____

   f. Which contacts are used to ensure that both motors cannot be energized at the same time?

      ____

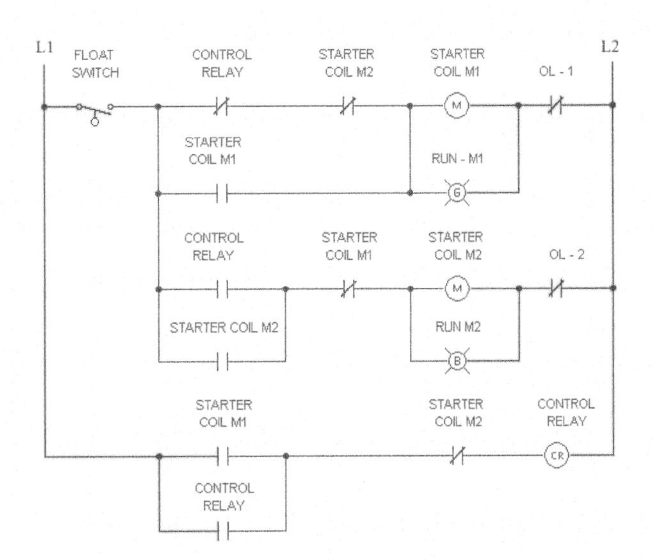

Alternating pumping operation control circuit.

TROUBLESHOOTING ASSIGNMENTS

1. This assignment involves the **troubleshooting** of **inoperative** break-make pushbutton circuit. Download the troubleshooting assignment circuit file **CH4 T01** from the **Diagrams-Troubleshooting-NEMA** folder.

   a. Turn on the Main Power Switch. Close the main CB and describe the faulty operating condition that exists ______________________________

   ____

   ____

   b. With the main CB closed, use the test probe Power function to record the (**ON/OFF**) power state of the following test points:

   i. Red PL with break-make switch not activated. ____________

   ii. Red PL with break-make switch activated. ____________

   iii. Green PL with break-make switch not activated. ____________

   iv. Green PL with break-make switch activated. ____________

   c. Open the main CB and use the test probe Continuity function to record the (**Yes/No**) continuity state of the following test points:

   i. Across the PB, not activated, normally closed contacts. ____________

   ii. Across the PB, activated normally, closed contacts. ____________

   iv. Across the PB, not activated normally, open contacts. ____________

   v. Across the PB, activated normally, open contacts. ____________

   d. Based on your readings, what is the most likely problem?

   ____

   ____

   e. What specific power and continuity readings lead you to conclude this?

   ____

   ____

   (Turn the Main Power Switch off, replace the defective component, and operate the circuit to verify your answer.)

2. This assignment involves the **troubleshooting** of **inoperative** emergency stop pushbutton circuit. Download the troubleshooting assignment circuit file **CH4 T02** from the **Diagrams-Troubleshooting-NEMA** folder.

   a. Turn on the Main Power Switch. Close the main breaker and use the test probe Power function to record the (ON/OFF) power state of the following test points:

      i. X1 to X2. _____________

      ii. X2 to both sides of F3 fuse. _____________

      iii. X2 to both sides of the E-stop pushbutton. _____________

      iv. X2 to both sides of the Stop pushbutton. _____________

      iv. X2 to both sides of Start pushbutton. _____________

   b. Based on your readings, what is the most likely problem?

      ____

      ____

   c. What specific power readings lead you to conclude this?

      ____

      ____

      (Turn the Main Power Switch off, replace the defective component, and operate the circuit to verify your answer.)

3. This assignment involves the **troubleshooting** of **faulty operative** motor starter circuit that contains monitoring pilot lights. Download the troubleshooting assignment circuit file **CH4 T03** from the **Diagrams-Troubleshooting-NEMA** folder.

   a. Turn on the Main Breaker Switch. Operate the circuit and summarize the faulty operating condition that exists.

      ____

      ____

      ____

      ____

   b. Based on your understanding how the circuit is supposed to operate what is the most likely fault? Why?

      ____

      ____

      ____

      (Turn the Main Power Switch off, replace the defective component, and operate the circuit to verify your answer.)

4. This assignment involves the **troubleshooting** of **faulty operative** motor starter circuit that contains a monitoring push-to-test pilot light. Download the troubleshooting assignment circuit file **CH4 T04** from the **Diagrams-Troubleshooting-NEMA** folder.
   a. Turn on the Main Breaker Switch. Operate the circuit and outline the faulty operating condition that exists.

      ____

      ____

   b. Open the main circuit breaker and using the test probe Continuity function identify the fault condition that exists.

      ____

      (Turn the Main Power Switch off, replace the defective component, and operate the circuit to verify your answer.)

5. This assignment involves the **troubleshooting** of **faulty operative** motor pumping circuit with a three-position selector switch control circuit. Download the troubleshooting assignment circuit file **CH4 T05** from the **Diagrams-Troubleshooting-NEMA** folder.

   a. Turn on the Main Breaker Switch. Operate the circuit. Which operating mode is at fault?

   ____

   b. From your understanding of how the circuit is supposed to operate, what are the most two likely components that may be at fault?

   ____

   ____

   ____

   c. Open the main circuit breaker and using the test probe Continuity function determine which of these two components is at fault. Why?

   ____

   ____

   ____

   ____

   (Turn the Main Power Switch off, replace the defective component, and operate the circuit to verify your answer.)

6. This assignment involves the **troubleshooting** of a **faulty operative** two-pole limit switch circuit. Download the troubleshooting assignment circuit file **CH4 T06** from the **Diagrams-Troubleshooting-NEMA** folder.

   a. Turn on the Main Breaker Switch. Operate the circuit. When the limit switch is operated between the activated and not activated positions what operating problem is evident?

   ____

   ____

   b. What component problem most likely exists?

   ____

   c. What test would you carry out with the test probe to confirm this?

   ____

   ____

   (Turn the Main Power Switch off, replace the defective component, and operate the circuit to verify your answer.)

7. This assignment involves the **troubleshooting** of a **faulty operative** overtravel limit switch control circuit. Download the troubleshooting assignment circuit file **CH4 T07** from the **Diagrams-Troubleshooting-NEMA** folder.

   a. Turn on the Main Breaker Switch. Operate the circuit and outline the faulty operating condition that exists.

   ____

   ____

   b. With the main breaker open, use the test probe Continuity function to record the (YES/NO) continuity state of the following test points:

      i. Across the stop button (button not activated). ________________

      ii. Across the stop button (button activated). ________________

      iii. Across the auxiliary normally closed F contact. ________________

      iv. Across the normally closed reverse LS contact. ________________

   c. Based on your readings, what is the most likely problem?

   ____

   ____

   (Turn the Main Power Switch off, replace the defective component, and operate the circuit to verify your answer.)

8. This assignment involves the **troubleshooting** of a **faulty operative** Temperature motor control circuit. Download the troubleshooting assignment circuit file **CH4 T08** from the **Diagrams-Troubleshooting-NEMA** folder.
   a. Close the Manual Starter breaker and operate the circuit. Which of the two operating modes is at fault? ________________
   b. Set the selector switch to the operating position that is not working. Set the Manual starter breaker closed and the temperature switch in the closed position. Use the test probe Power function to record the (ON/OFF) power state of the following test points:
      i. T2 to T1. ____________
      ii. T2 to both side of contact "A."

      ____

      iii. T2 to both sides of the closed temperature switch.

      ____

      iv. Temperature switch to both sides of the terminal connection.

      ____

   c. Based on your readings, what is the most likely problem?

   ____

   ____

   (Turn the Main Power Switch off, replace the defective component, and operate the circuit to verify your answer.)

9. This assignment involves the **troubleshooting** of **inoperative** interposing relay motor control circuit. Download the troubleshooting assignment circuit file **CH4 T09** from the **Diagrams-Troubleshooting-NEMA** folder.
   a. Turn on the Main Power Switch. Close the main breaker and use the test probe Power function to record the (ON/OFF) power state, in sequence, of the following **control circuit** test points:
      i. H1 to H4. ________________
      ii. X1 to X2. ________________
      iii. X2 to both sides of F3 fuse. ________________
      iv. X2 to both sides of the closed ON/OFF switch. ________________
      v. X2 to both sides of the closed proximity switch. ________________
      vi. Across relay coil CR. ____________________
   b. Continue to use the test probe Power function (ON/OFF and Proximity switches closed) to record the power state of the following **power circuit** test points:
      i. Across Starter coil M. ________________
      ii. Line side of the M power contacts (all three leads). ___________
      iii. Load side of the M power contacts (all three leads). ___________
   c. Based on your readings, what is the most likely problem?

      ____

      ____

      ____

      (Turn the Main Power Switch off, replace the defective component, and operate the circuit to verify your answer.)

10. This assignment involves the **troubleshooting** of a **faulty operative** tanking filling and emptying motor control circuit. Download the troubleshooting assignment circuit file **CH4 T10** from the **Diagrams-Troubleshooting-NEMA** folder.

    a. Turn on the Main Power Switch. The operator of this process reports the problem as follows:

        - With the liquid level of the tank below the empty level mark, momentarily closing the FILL button starts the tank filling with liquid.
        - When the liquid reaches the full level, the process operates normally to automatically stop the liquid flow.
        - With the tank full, nothing happens when the EMPTY button is activated.

    Close the main breaker and use the test probe Power function to record the (ON/OFF) power state, of the following circuit test points:

        i. L2 to the empty button (sensor side-button not activated). ___________
        ii. L2 to the empty button (sensor side-button activated). ______________
        iii. Across coil 2CR (empty button activated-sensor switch open). _______
        iv. Across coil 2CR (empty button activated-sensor switch closed). ______
        v. Across solenoid B (empty button activated-sensor switch closed). ____

    b. Based on your readings, what is the most likely problem?

    ____

    ____

    c. Turn the Main Power switch off, replace the defective component, and repeat your readings:

        i. L2 to the empty button (sensor side-button not activated). ___________
        ii. L2 to the empty button (sensor side-button activated). ______________
        iii. Across coil 2CR (empty button activated-sensor switch open). _______
        iv. Across coil 2CR (empty button activated-sensor switch closed). ______
        v. Across solenoid B (empty button activated-sensor switch closed). ____

11. This assignment involves the **troubleshooting** of a **faulty operative** alternating pumping motor control circuit. Download the troubleshooting assignment circuit file **CH4 T11** from the **Diagrams-Troubleshooting-NEMA** folder.

    a. Close the Manual Starter and Circuit breakers. Operate the circuit by initially closing the normally open float switch. What malfunction, if any appears to occur?

    ____

    b. Open the float switch. What malfunction, if any, appears to occur?

    ____

    ____

    c. Close the float switch. What malfunction, if any, appears to occur?

    ____

    ____

    d. At this point, use the test probe Power function to record the (ON/OFF) power state of the following control circuit test points:

        i. L2 to Starter coil M1 contact (both sides). ________________

        ii. L2 to Starter coil M2-NC contact (both sides). ________________

        iii. Across control relay coil CR. ________________

    e. Based on your readings, what is the most likely problem? Why?

    ____

    (Turn the Main Power Switch off, replace the defective component, and operate the circuit to verify your answer.)

5 Electric Motors

CIRCUIT ANALYSIS ASSIGNMENTS

1. The purpose of this assignment is to analyze the operation of the **DC reversing motor starter** used to operate a shunt motor in the forward and reverse directions shown. Download the analysis assignment circuit file **CH5 Assign 01** from the **Diagrams-Analysis-NEMA** folder. Click on the Main Power Switch (ON/OFF) located on the left side of the toolbar. Left-click on the circuit breaker to close it. Left-click on the directional push buttons to operate the motor in the forward and reverse directions.
   a. How is reversing of the motor accomplished?

   ____

   b. Assume the motor is running in the forward direction and the reverse push button is closed. What happens as a result? Why?

   ____

   ____

   c. Close the forward and reverse power contact by left-clicking on them. This will simulate a fault condition resulting from both the forward and reverse contactors being energized at the same time. Click the main circuit breaker on. What would happen? Why?

   ____

   ____

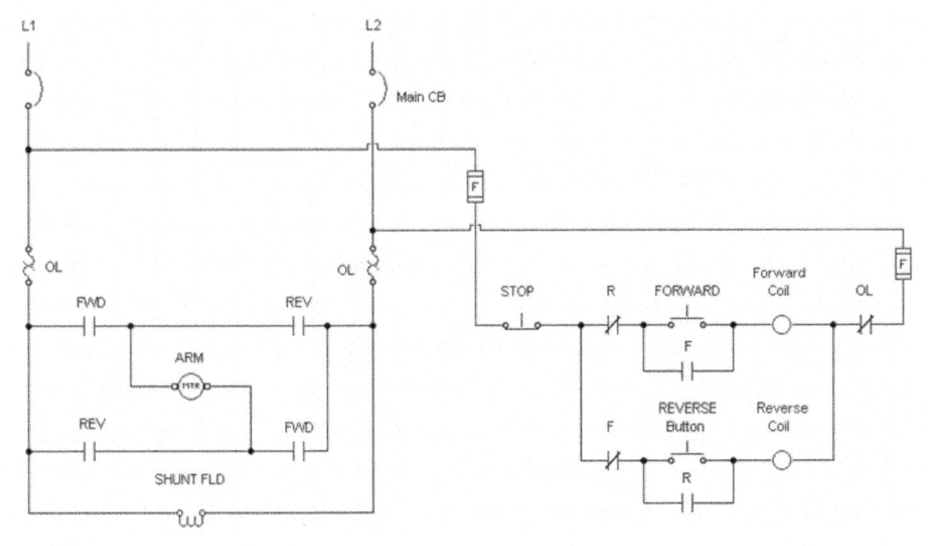

DC reversing motor starter.

2. The purpose of this assignment is to analyze the operation of the **AC reversing motor starter** used to operate a three-phase squirrel-cage motor in the forward and reverse directions shown. Download the analysis assignment circuit file **CH5 Assign 02** from the **Diagrams-Analysis-NEMA** folder. Click on the Main Power Switch (ON/OFF) located on the left side of the toolbar. Left-click on the circuit breaker to close it. Left-click on the directional push buttons to operate the motor in the forward and reverse directions.

   a. How is reversing of the motor accomplished?

   ____

   b. When the motor is operating in the forward direction, L1, L2, and L3 connect to motor leads ______________ respectively.

   c. When the motor is operating in the reverse direction, L1, L2, and L3 connect to motor leads ______________ respectively.

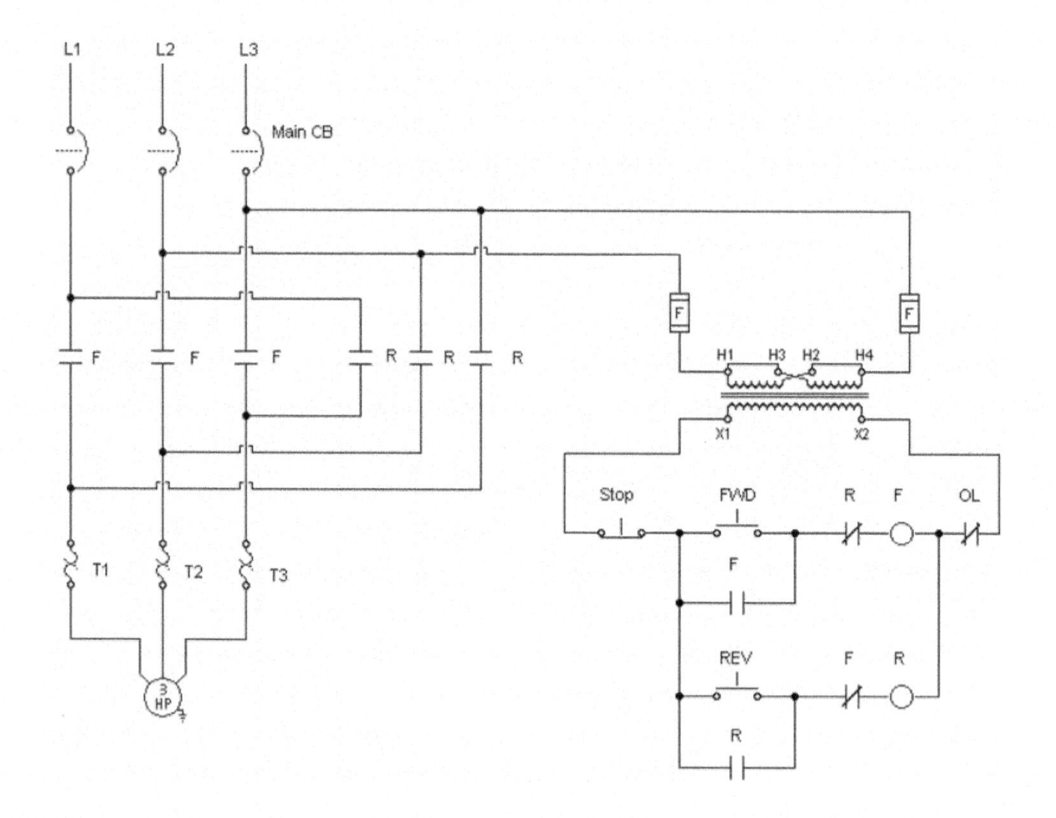

AC reversing motor starter.

TROUBLESHOOTING ASSIGNMENTS

1. This assignment involves the **troubleshooting** of a **faulty operative** DC reversing motor starter. Download the **troubleshooting** assignment circuit file **CH5 T01** from the **Diagrams-Troubleshooting-NEMA** folder.
   a. The motor operates in the forward direction only. Turn on the Main Power Switch. Close the main CB and use the test probe Power function to record the (ON/OFF) power state, in sequence, of the following control circuit test points:
      i. L1 to L2 on the load side of the control circuit fuses.

      ____

      ii. L2 to both sides of the stop button (button not activated).

      ____

      iii. L2 to both sides of the normally closed F contact.

      ____

      iv. L2 to load side of the reverse button (after being momentarily activated).

      ____

      v. Across the reverse coil. ______________________________

   b. With the circuit still in its last state, record the (ON/OFF) power state of the following power circuit test points:
      i. L1 to L2 on the load side of the OL heater coils. ____________
      ii. Across the motor Armature (ARM). ________________
   c. Based on your readings, what is the most likely problem? Why?

      ____

      ____

      ____

      (Turn the Main Power Switch off, replace the defective component, and operate the circuit to verify your answer.)

2. This assignment involves the **troubleshooting** of an **inoperative** AC reversing motor starter circuit. Download the troubleshooting assignment circuit file **CH5 T02** from the **Diagrams-Troubleshooting-NEMA** folder.

   a. The report from the operator of this process states that the motor fails to start when operated in the forward or reverse direction. When activated in either direction the motor hums but will not start. After a few seconds it automatically turns OFF and the humming stops. Turn on the Main Power Switch and close the main CB. Operate the circuit to verify this problem. What component most likely caused the motor circuit to automatically shut down after a few seconds of the startup?

      ____

   b. Open the main CB and use the test probe Continuity function to record the (Yes/No) continuity state of the following test points:

      i. Across OL heater coil for T1. ______________

      ii. Across OL heater coil for T2. ______________

      iii. Across OL heater coil for T3. ______________

   c. Based on your readings, what is the most likely problem?

      ____

      ____

      (Turn the Main Power Switch off, replace the defective component, and operate the circuit to verify your answer.)

6 Contactors and Motor Starters

CIRCUIT ANALYSIS ASSIGNMENTS

1. The purpose of this assignment is to analyze the operation of the **magnetic contactor control of a heater load** shown. Download the analysis assignment circuit file **CH6 Assign 01** from the **Diagrams-Analysis-NEMA** folder. Click on the Main Power Switch (ON/OFF) located on the left side of the toolbar. Left-click on the circuit breaker to close it. Left-click on the heater operating control devices to turn the heater on and off.

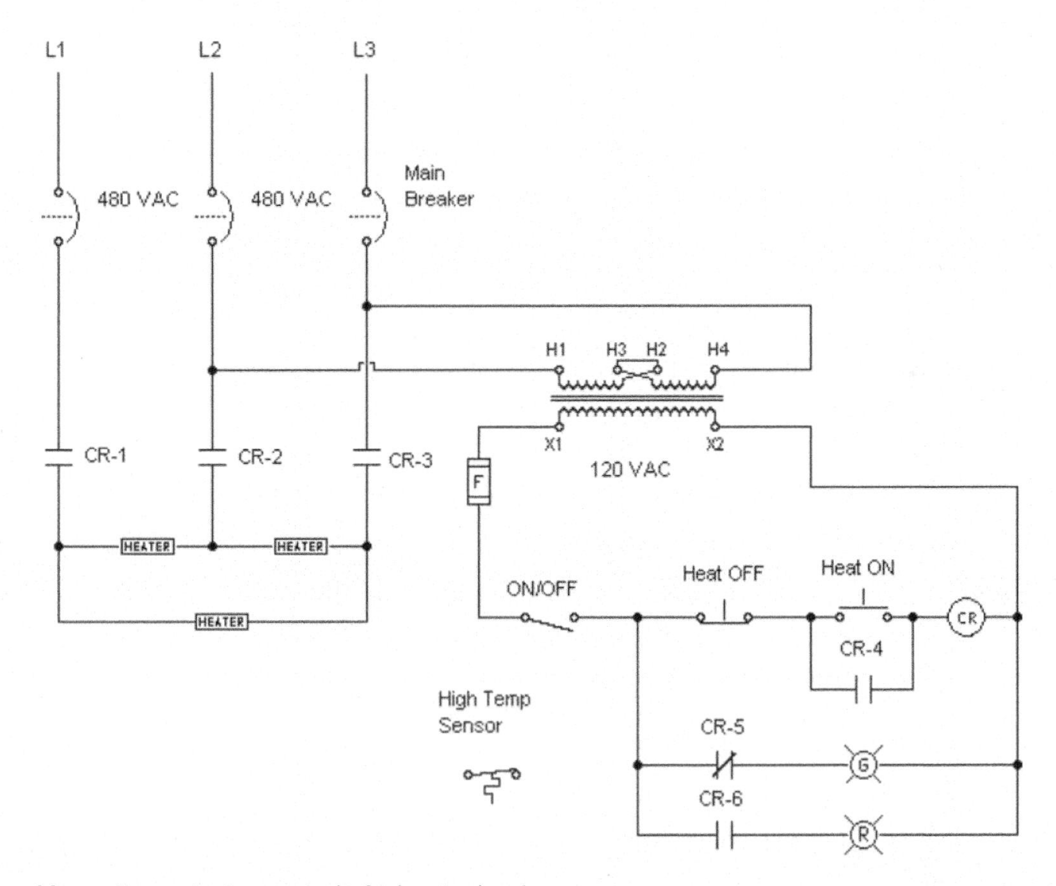

Magnetic contactor control of a heater load.

a. The value of the power circuit voltage is ______________ V.

b. The value of the control circuit voltage is ______________ V.

c. The value of the contactor coil voltage is ______________ V.

d. When on, what do the green and red pilot lights indicate?

____

____

e. Left-click one of the heaters in the three-phase bank to open-circuit it. In what way does this affect the control and power circuit?

____

____

f. Right-click on any two of the heaters in the three-phase bank to short-circuit them. In what way does this affect the control and power circuit?

____

____

g. What type of contacts would CR-4, CR-5, and CR-6 be classified as?

____

h. A high-temperature sensor, equipped with a single normally closed contact, is to be wired into the circuit to prevent the heaters from being energized should the temperature rise above a preset safe value. Explain how this sensor would be connected into the circuit. Switch to the normal mode and connect the temperature sensor into the circuit. Simulate the operation of the circuit with the sensor connected.

____

____

2. The purpose of this assignment is to analyze the operation of the **dual-coil mechanically held lighting contactor** shown. Download the analysis assignment circuit file **CH6 Assign 02** from the **Diagrams-Analysis-NEMA** folder. Click on the Main Power Switch (ON/OFF) located on the left side of the toolbar. Left-click on the main circuit breaker. Turn the branch circuit light banks on and off.

a. The two lighting circuits are ______________-phase and generally rated at ______________ V or ______________ V.

b. With the main circuit breaker on, what procedure is followed to turn both banks of lights on?

____

c. Whenever both banks of lights are on, what is the state (open/closed) of contacts M3 and M4?

____

d. Whenever both banks of lights are on, what is the value of the current flow through the contactor coil?

____

e. To what points does the common lead of the latch/unlatch coils connect?

____

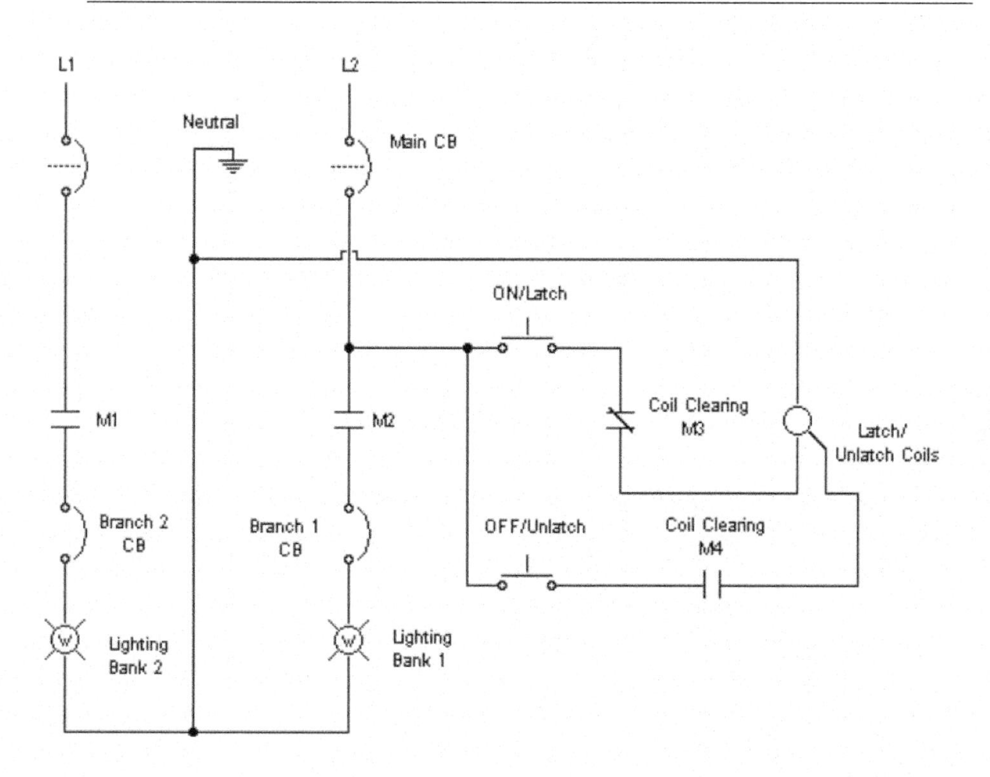

Dual-coil mechanically held lighting contactor.

TROUBLESHOOTING ASSIGNMENTS

1. This assignment involves the **troubleshooting** of a **faulty operative** contactor control of a heater load. Download the troubleshooting assignment circuit file **CH6 T01** from the **Diagrams-Troubleshooting-NEMA** folder.

   a. Turn on the Main Power Switch. Close the main breaker and ON/OFF switch. Which pilot light comes on? Why?

   ____

   ____

   ____

   b. Next, momentarily close the Heat ON pushbutton. What problem is evident by reaction of the two pilot lights?

   ____

   ____

   c. Open the Main Breaker and use the test probe Continuity function to record the (YES/NO) continuity state of the following control circuit test points:

      i. Across the Heat OFF pushbutton (not activated). ___________

      ii. Across the Heat OFF pushbutton (activated). ______________

      iii. Across the Heat ON pushbutton (not activated). ___________

      iv. Across the Heat ON pushbutton (activated). ______________

      v. Across the High Temp Sensor (not activated). ______________

      vi. Across the High Temp Sensor (activated). ______________

   d. Based on your readings, what is the most likely problem? Why?

   ____

   ____

   (Turn the Main Power Switch off, replace the defective component, and operate the circuit to verify your answer.)

2. This assignment involves the **troubleshooting** of a **faulty operative** dual-coil mechanically held lighting contactor. Download the troubleshooting assignment circuit file **CH6 T02** from the from the **Diagrams-Troubleshooting-NEMA** folder.

a. Turn on the Main CB. Close the main breaker and ON/OFF switch. Operate the circuit and describe the faulty condition that exists.

____

____

____

b. A fault in the control circuit is suspected. Open the main CB and use the test probe function to record the (YES/NO) continuity state of the following test points.

   i. Between the Neutral and the Latch/Unlatch Coil. ___________

   ii. Between the Latch/Unlatch Coil and the M4 contact. _____________

   iii. Between the M4 contact and the OFF/Unlatch pushbutton. __________

   iv. Between the OFF/Unlatch and the ON/Latch pushbutton. ___________

c. Based on your readings, what is the most likely problem?

____

____

(Turn the Main Power Switch off, replace the defective component, and operate the circuit to verify your answer.)

7 Relays

CIRCUIT ANALYSIS ASSIGNMENTS

1. The purpose of this assignment is to analyze the **multiple switching operation of the relay** shown. Download the analysis assignment circuit file **CH7 Assign 01** from the **Diagrams-Analysis-NEMA** folder. Click on the Main Power Switch (ON/OFF) located on the left side of the toolbar. Left-click on the circuit breaker to close it and on the push button to operate.
   a. Record the state of each light with the push button open.
      R - ______________, G - ______________, Y - ______________
   b. Record the state of each light with the push button closed.
      R - ______________, G - ______________, Y - ______________
   c. Which light remains on with the push button open or closed? Why?
      ____
      ____
   d. Which light(s) is/are on when the relay coil is energized?
      ____

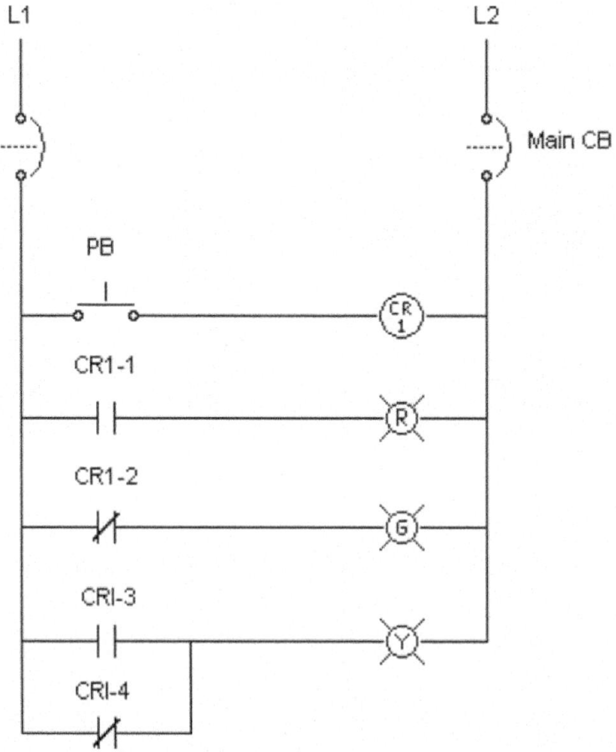

Multiple switching relay circuit.

2. The purpose of this assignment is to analyze the **operation of the instantaneous and delayed switching contacts** of the **relay** shown. Download the analysis assignment circuit file **CH7 Assign 02** from the **Diagrams-Analysis-NEMA** folder. Click on the Main Power Switch (ON/OFF) located on the left side of the toolbar. Left-click on the circuit breaker to close it and on the switch to operate it.

   a. Record the state of each light with the switch open.

      R - __________, G - __________, Y - __________, W- __________

   b. Record the state of each light immediately upon closing the switch.

      R - __________, G - __________, Y - __________, W- __________

   c. Record the state of each light 6 seconds after the switch has been closed.

      R - __________, G - __________, Y - __________, W- __________

   d. Record the state of each light immediately upon opening the switch.

      R - __________, G - __________, Y - __________, W- __________

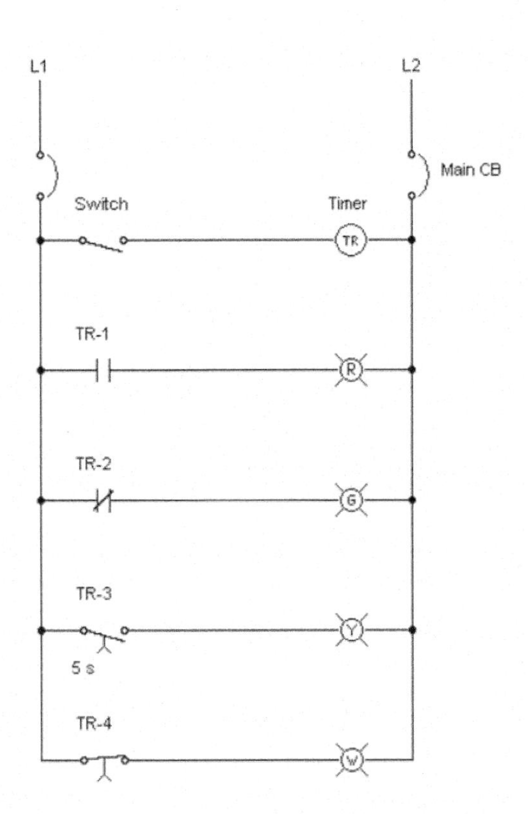

Instantaneous and delayed switching contacts.

3. The purpose of this assignment is to analyze the **switching operation of the on-delay timer** shown. Download the analysis assignment circuit file **CH7 Assign 03** from the **Diagrams-Analysis-NEMA** folder. Click on the Main Power Switch (ON/OFF) located on the left side of the toolbar. Left-click on the circuit breaker to close it and on the switch to operate it.

   a. Record the state of each light with the switch open.

      R - _______________, G - _______________

   b. Record the state of each light immediately upon closing the switch.

      R - _______________, G - _______________

   c. Record the state of each light 10 seconds after the switch has been closed.

      R - _______________, G - _______________

   d. Record the state of each light immediately upon opening the switch.

      R - _______________, G - _______________

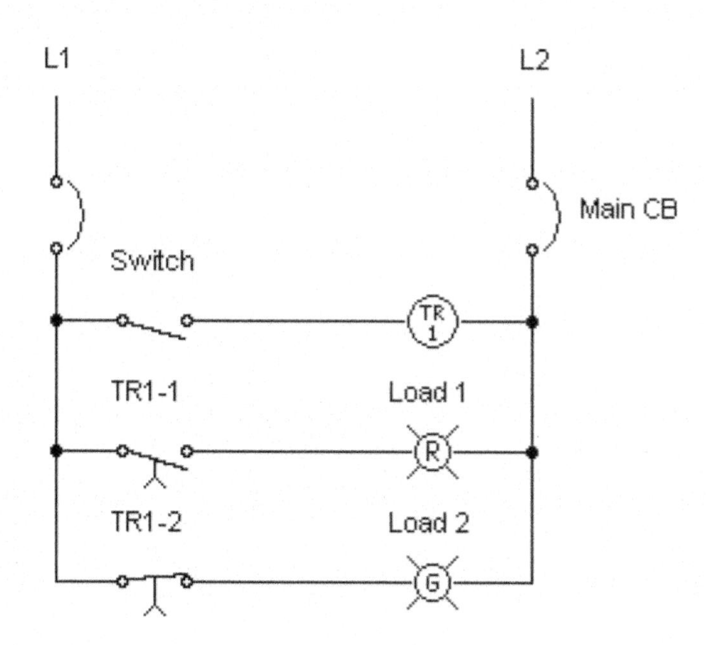

On-delay timer circuit.

4. The purpose of this assignment is to analyze the **switching operation of the off-delay timer** shown. Download the analysis assignment circuit file **CH7 Assign 04** from the **Diagrams-Analysis-NEMA** folder. Click on the Main Power Switch (ON/OFF) located on the left side of the toolbar. Left-click on the circuit breaker to close it and on the switch to operate it.

   a. Record the state of each light with the switch open.

      R - ________________, G - ________________

   b. Record the state of each light immediately upon closing the switch.

      R - ________________, G - ________________

   c. Record the state of each light 10 seconds after the switch has been closed.

      R - ________________, G - ________________

   d. Record the state of each light immediately upon opening the switch.

      R - ________________, G - ________________

   e. Record the state of each light 10 seconds after opening the switch.

      R - ________________, G - ________________

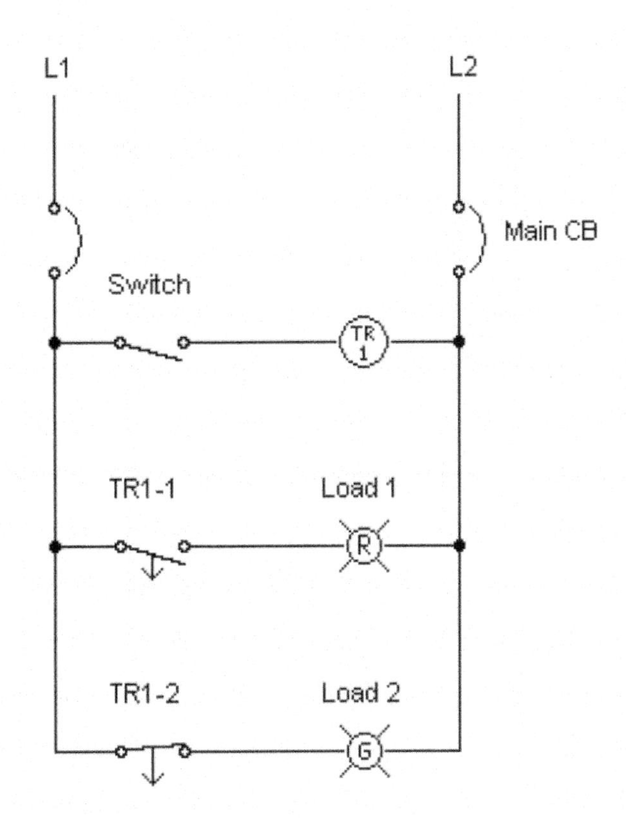

Off-delay timer circuit.

5. The purpose of this assignment is to analyze the **switching operation of the recycle timer** shown. Download the analysis assignment circuit file **CH7 Assign 05** from the **Diagrams-Analysis-NEMA** folder. Click on the Main Power Switch (ON/OFF) located on the left side of the toolbar. Left-click on the circuit breaker to close it and on the switch to operate it.

   a. What type of electrical device is the circuit functioning as?

   ____

   b. What is the value of the on time delay period?

   ____

   c. What is the value of the off time delay period?

   ____

   d. How is the recycling sequence turned on and off?

   ____

   e. What type of symbol is used for the timed contact?

   ____

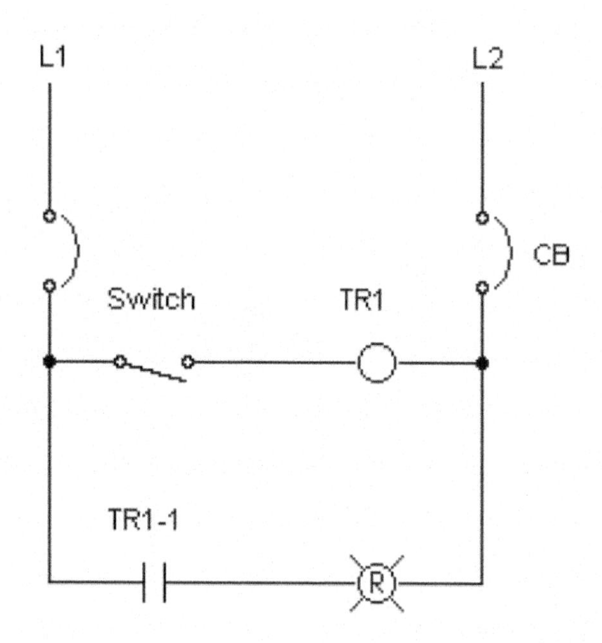

Recycle timer circuit.

6. The purpose of this assignment is to analyze the **switching operation of the one-shot timer** shown. Download the analysis assignment csircuit file **CH7 Assign 06** from the **Diagrams-Analysis-NEMA** folder. Click on the Main Power Switch (ON/OFF) located on the left side of the toolbar. Left-click on the circuit breaker to close it and on the push button to operate it.
   a. What type of timer (on- or off-delay) is TR1 and TR2?

      ____

   b. Outline the sequence of operations that occur when power is first applied and the push button is held continuously closed.

      ____

      ____

      ____

   c. Which timer setting determines how long the light remains on?

      ____

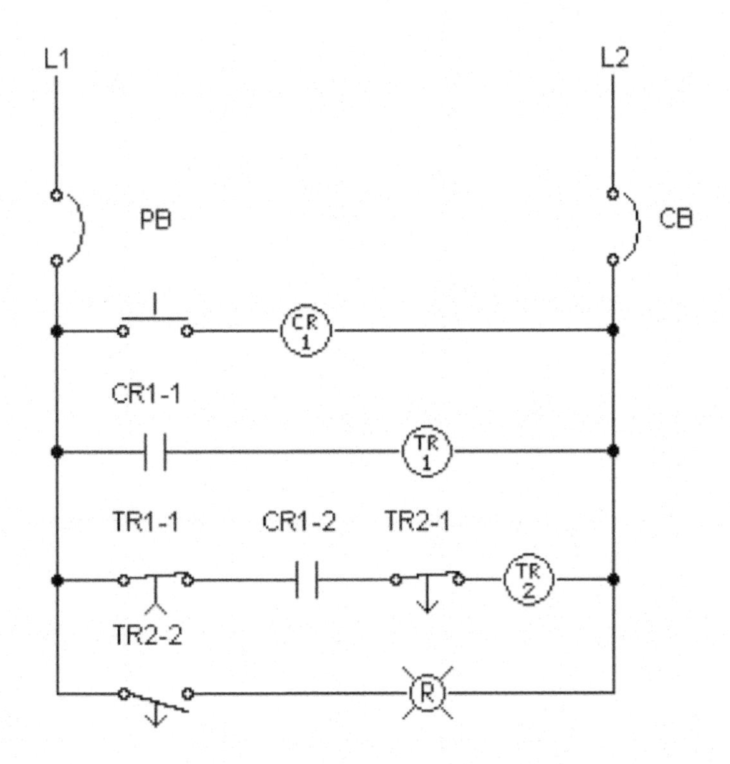

One-shot timer circuit.

7. The purpose of this assignment is to analyze the **operation of the latching relay** shown. Download the analysis assignment circuit file **CH7 Assign 07** from the **Diagrams-Analysis-NEMA** folder. Click on the Main Power Switch (ON/OFF) located on the left side of the toolbar. Left-click on the circuit breaker to close it and on the push buttons to operate them. This latching relay is configured using a dual coil with separate set (latch) and reset (unlatch) leads.

   a. Momentarily close the latch button and make a note of how the circuit reacts.

   Answer ____

   b. Momentarily close the unlatch button and make a note of how the circuit reacts.

   Answer ____

   c. With the pilot light on, open and close the main circuit breaker and make a note of how the circuit reacts.

   Answer ____

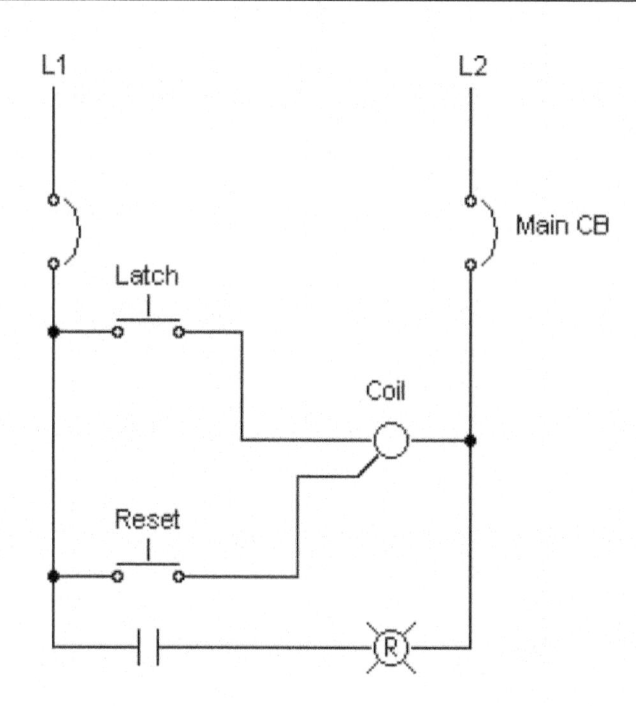

Latching relay circuit.

8. The purpose of this assignment is to analyze the **operation of the alternating relay** shown. Download the analysis assignment circuit file **CH7 Assign 08** from the **Diagrams-Analysis-NEMA** folder. Click on the Main Power Switch (ON/OFF) located on the left side of the toolbar. Left-click on the circuit breaker to close it and on the push button to operate it. Record the status of each pilot light when the following operations are preformed in sequence.

   a. Power is first applied.

   Answer ______________________________

   b. Push button is pressed and released.

   Answer ______________________________

   c. Push button is pressed and released a second time.

   Answer ______________________________

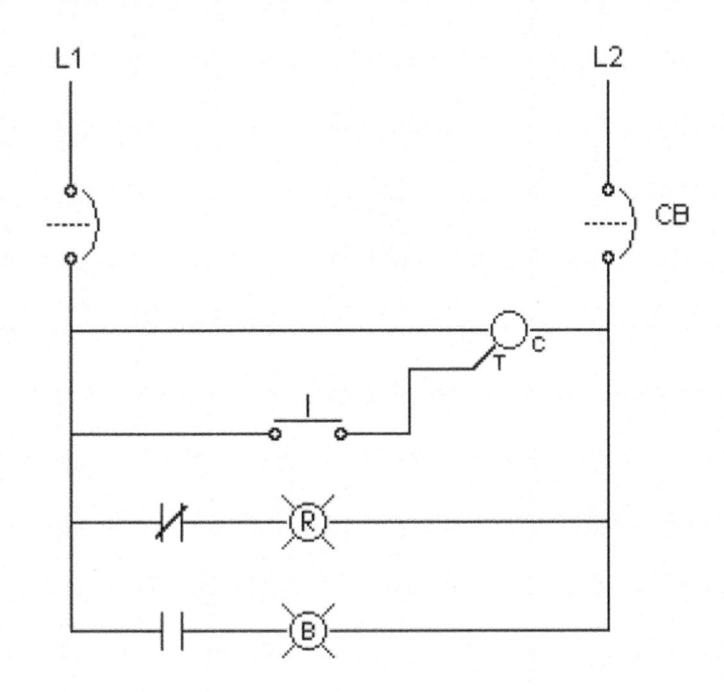

Basic alternating relay circuit.

9. The purpose of this assignment is to analyze the **operation of an alternating relay as part of a motor control circuit** shown. Download the analysis assignment circuit file **CH7 Assign 09** from the **Diagrams-Analysis-NEMA** folder. Click on the Main Power Switch (ON/OFF) located on the left side of the toolbar. Left-click on the circuit breaker to close it and on the push buttons to operate them.

   a. State the purpose of the alternating relay used in the circuit

   Answer ____

   ____

   b. Outline the sequence of operations.

   Answer ____

   ____

   ____

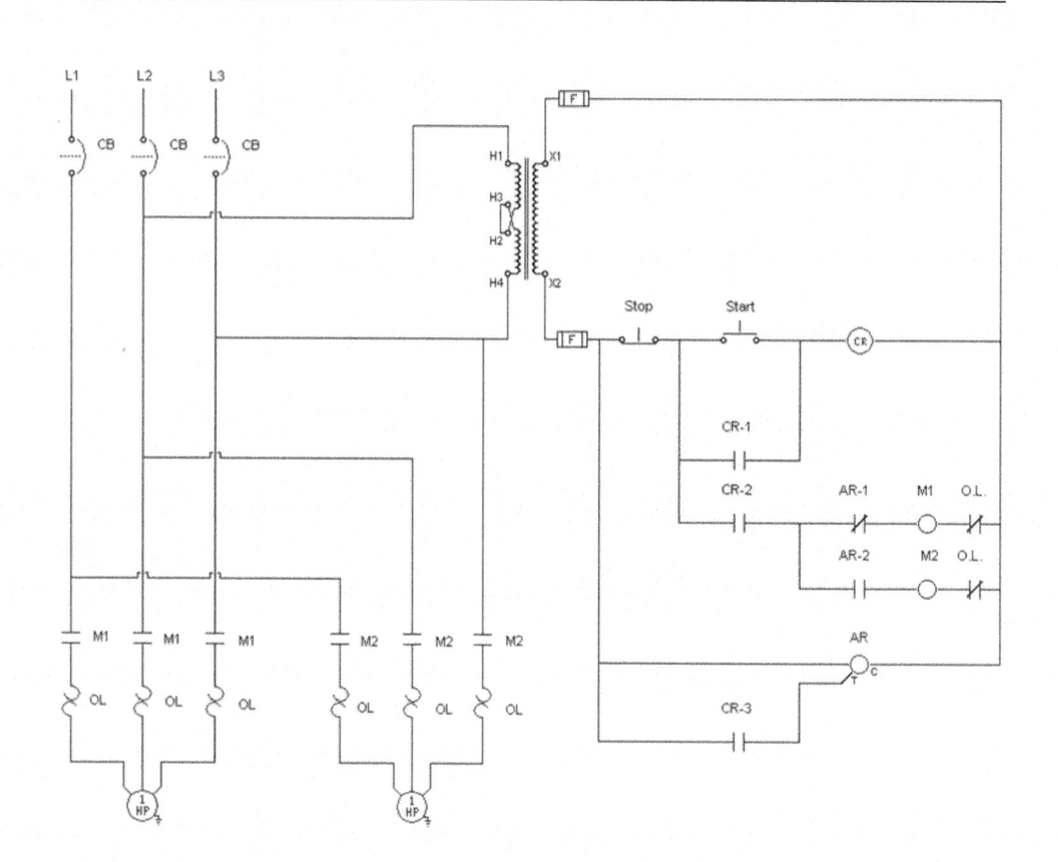

Alternating relay as part of a motor control circuit.

10. The purpose of this assignment is to analyze **the relay control logic functions** shown. Download the analysis assignment circuit file **CH7 Assign 10** from the **Diagrams-Analysis-NEMA** folder. Click on the Main Power Switch (ON/OFF) located on the left side of the toolbar. Left-click on the circuit breaker to close it and on the input control devices to operate the loads. Identify the type of logic function associated with each of the control rungs A to E.

    a. Rung A - _______________ b. Rung B - _______________

    c. Rung C - _______________ d. Rung D - _______________

    e. Rung E - _______________

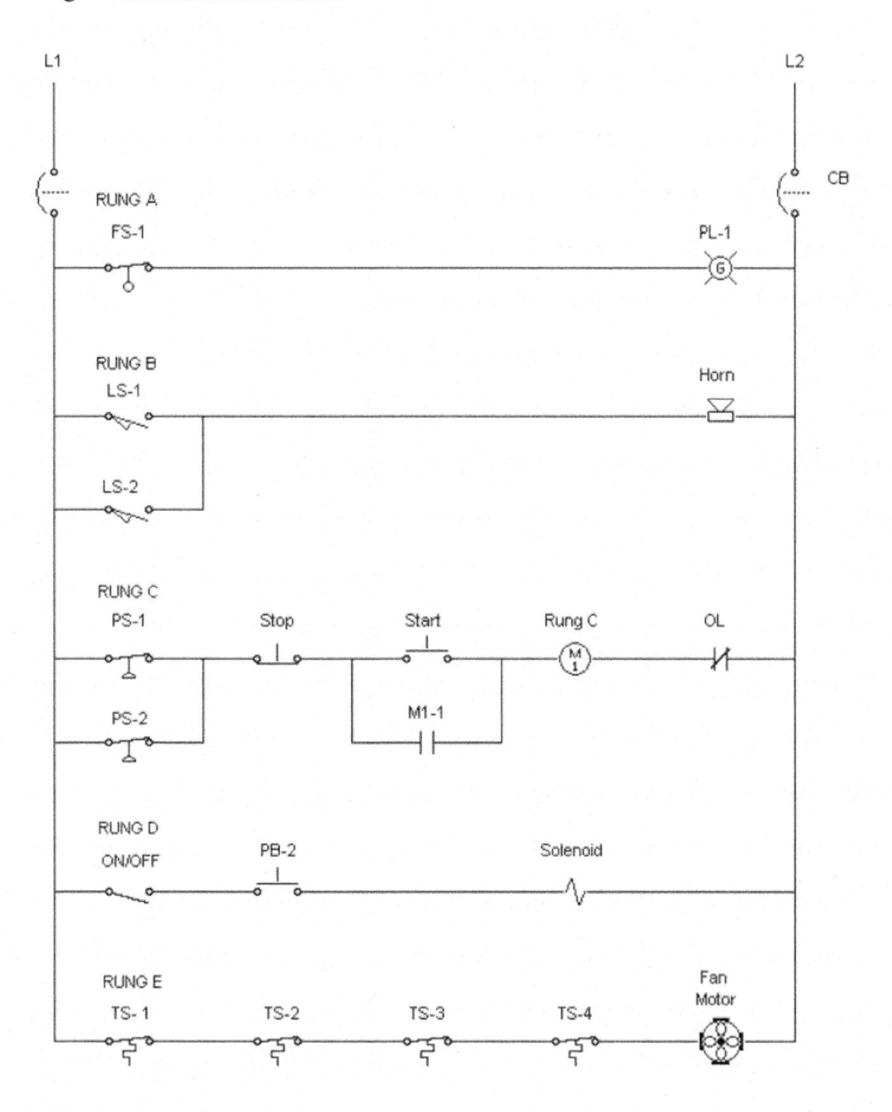

Relay control logic functions.

11. The purpose of this assignment is to analyze **troubleshooting by analysis of the circuit logic** shown. Download the analysis assignment circuit file **CH7 Assign 11** from the **Diagrams-Analysis-NEMA** folder. Click on the Main Power Switch (ON/OFF) located on the left side of the toolbar. Left-click on the circuit breaker to close it and on the input control devices to operate the loads. Operate the circuit and respond to each of the following troubleshooting scenarios.

    a. Assume the pressure switch is stuck in the open position. How would this affect the operation of the pilot light?

    Answer ____

    ____

    b. Right-click on the solenoid to fault it shorted and operate the limit switches. What happens? Why?

    Answer ____

    ____

    c. Assume proximity switch 2 is fault shorted. How would this affect the operation of the motor starter circuit?

    Answer ____

    ____

    d. Assume the temperature switch is faulted open. How would this affect the operation of the fan motor?

    Answer ____

    ____

    e. Assume that during normal operation and with the fluorescent light bank on, PB3 becomes fault shorted. How would this affect the operation of the light circuit?

    Answer ____

    ____

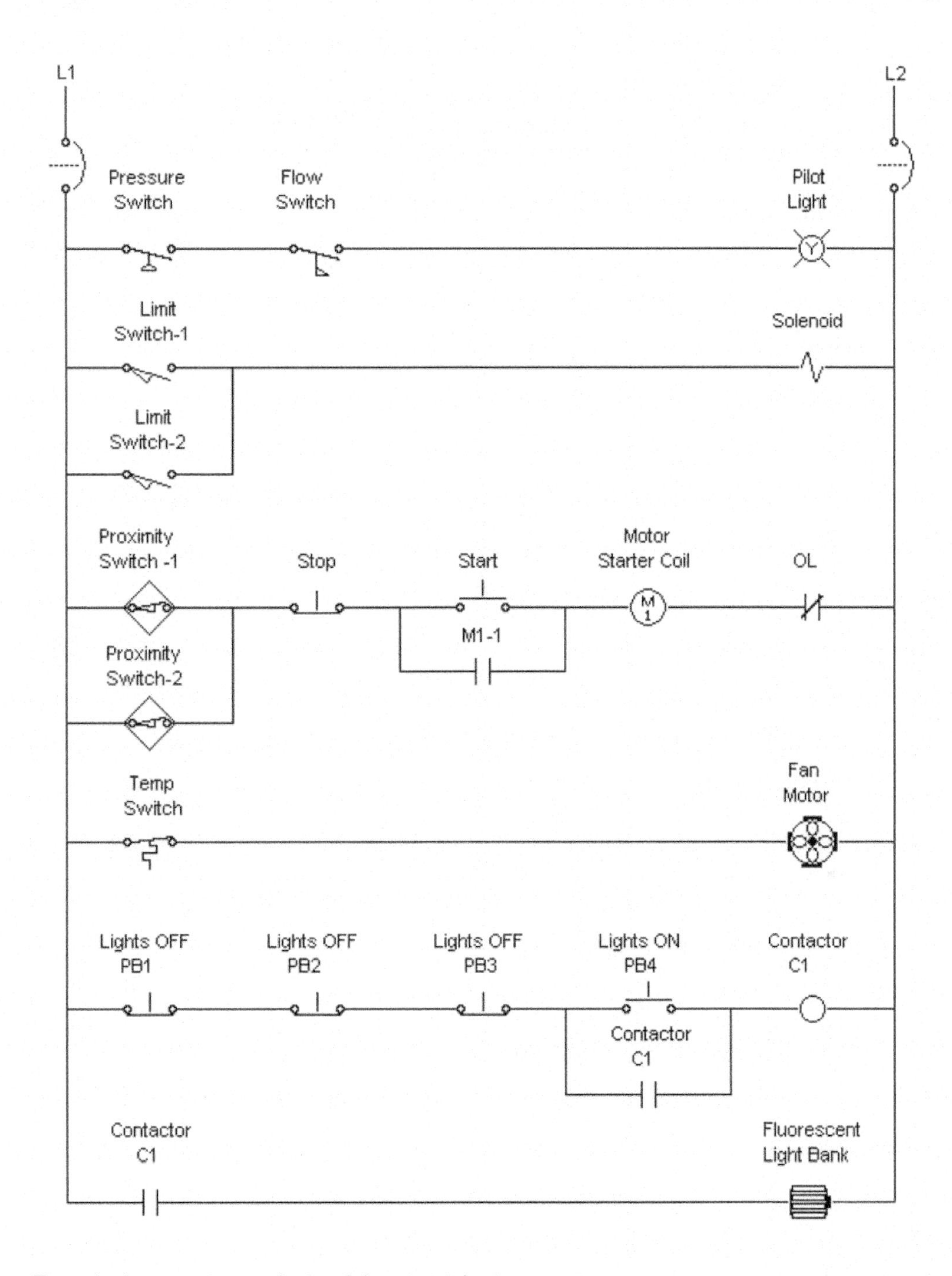

Troubleshooting by analysis of the circuit logic.

12. The purpose of this assignment is to analyze the **application of the interposing relay current control circuit** shown. Download the analysis assignment circuit file **CH7 Assign 12** from the **Diagrams-Analysis-NEMA** folder. Click on the Main Power Switch (ON/OFF) and Main Breaker to turn on the simulation.

   a. Compare the value of the power circuit and control circuit operating voltages.

   ____

   b. Right-click the TEMP Switch to close it. Summarize the circuit changes that take.

   ____

   ____

   ____

   c. Right-click the TEMP Switch to open it. Summarize the circuit changes that take place.

   ____

   ____

   ____

   d. Taking into consideration that this a 10HP motor load, why would the Relay coil current requirement be much lower than that of the Starter coil?

   ____

   ____

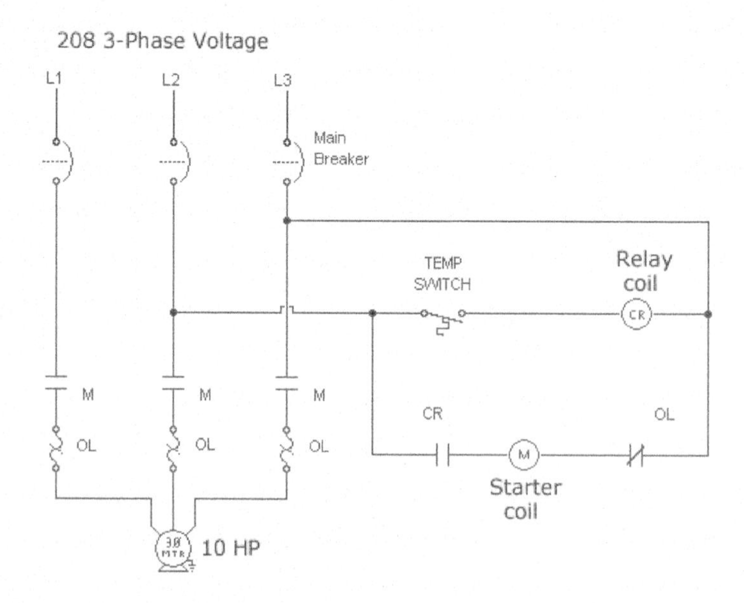

Interposing relay current control circuit.

13. The purpose of this assignment is to analyze the **application of the interposing relay voltage control circuit** shown. Download the analysis assignment circuit file **CH7 Assign 13** from the **Diagrams-Analysis-NEMA** folder. Click on the Main Power Switch (ON/OFF) and Main Breaker to turn on the simulation.

    a. Compare the value of the power circuit and control circuit operating voltages.

    ____

    b. Right-click the TEMP Switch to close it. Summarize the circuit changes that take.

    ____

    ____

    ____

    c. Right-click the TEMP Switch to open it. Summarize the circuit changes that take place.

    ____

    ____

    ____

    d. Taking into consideration the differences between the power and control circuit voltages, what is the advantage of using the TEMP SWITCH with a the lower operating voltage circuit?

    ____

    ____

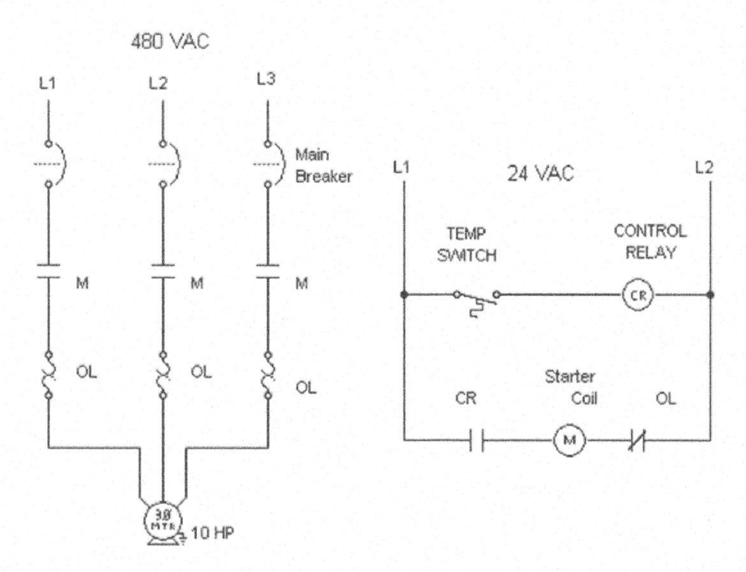

Interposing relay voltage control circuit.

14. The purpose of this assignment is to analyze the **application of the on-delay timer warning signal** shown. Download the analysis assignment circuit file **CH7 Assign 14** from the **Diagrams-Analysis-NEMA** folder. Click on the Main Power Switch (ON/OFF) and Main Switch to turn on the simulation.

    a. Momentarily close the Start button and record the resultant state (ON/OFF or Energized/Deenergized or Open/Closed) of each of the following:

    i. CR coil __________ ii. CR contact __________ iii. TR coil __________

    iv. Alarm __________ v. TR contact __________ vi. M coil __________

    vii. M (NO) contacts __________ viii. M (NC) contacts __________

    b. Once the timer has timed out, record the resultant state of each of the following:

    i. CR coil __________ ii. CR contact __________ iii. TR coil __________

    iv. Alarm __________ v. TR contact __________ vi. M coil __________

    vii. M (NO) contacts __________ viii. M (NC) contacts __________

    c. What is the time of the ON time delay period? ______________

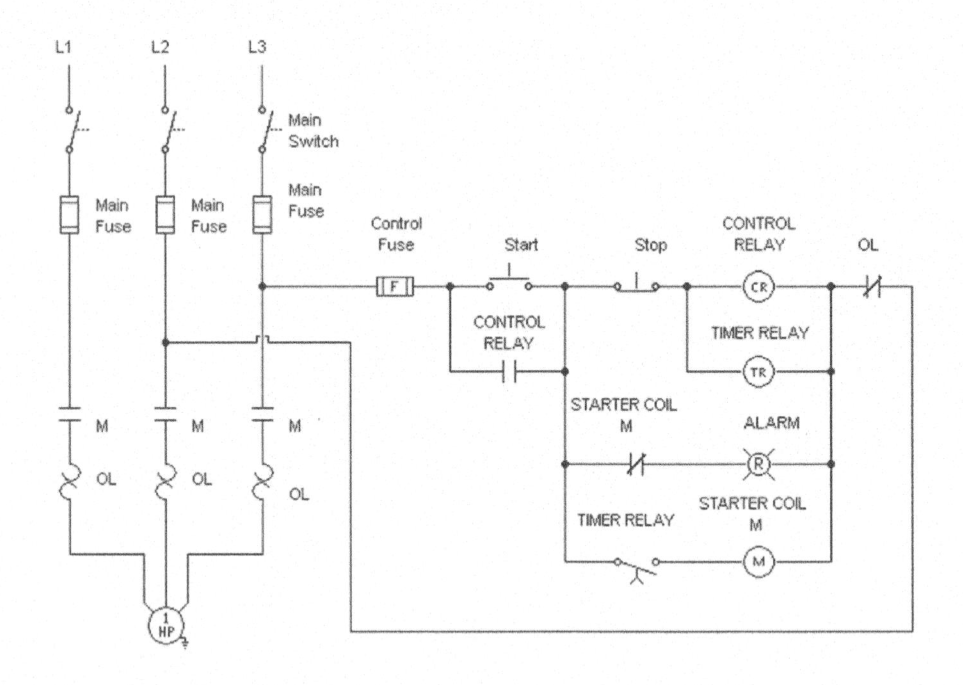

On-delay timer warning signal.

15. The purpose of this assignment is to analyze the **application of the off-delay timer instantaneous and delayed switching circuit** shown. Download the analysis assignment circuit file **CH7 Assign 15** from the **Diagrams-Analysis-NEMA** folder. Click on the Main CB to turn on the simulation. When power is first applied and LS1 open, record the resultant state (ON/OFF or Energized/Deenergized or Open/Closed) of each of the following:

a. i. TR coil ________ ii. TR-1 contact ________ iii. TR-2 contact ________
iv. TR-3 contact ________ v. TR-4 contact ________ vi. M1 coil ________
vii. M2 coil ________ viii. G light ________ ix. R light ________

b. When LS1 is then closed, record the resultant state (ON/OFF or Energized/Deenergized or Open/Closed) of each of the following:
i. TR coil ________ ii. TR-1 contact ________ iii. TR-2 contact ________
iv. TR-3 contact ________ v. TR-4 contact ________ vi. M1 coil ________
vii. M2 coil ________ viii. G light ________ ix. R light ________

c. When LS1 is then opened to begin the timing period, record the resultant state (ON/OFF or Energized/Deenergized or Open/Closed) of each of the following:
i. TR coil ________ ii. TR-1 contact ________ iii. TR-2 contact ________
iv. TR-3 contact ________ v. TR-4 contact ________ vi. M1 coil ________
vii. M2 coil ________ viii. G light ________ ix. R light ________

d. After the timing period has lapsed, record the resultant state (ON/OFF or Energized/Deenergized or Open/Closed) of each of the following:
i. TR coil ________ ii. TR-1 contact ________ iii. TR-2 contact ________
iv. TR-3 contact ________ v. TR-4 contact ________ vi. M1 coil ________
vii. M2 coil ________ viii. G light ________ ix. R light ________

e. What is the time of the OFF time delay period? ____________

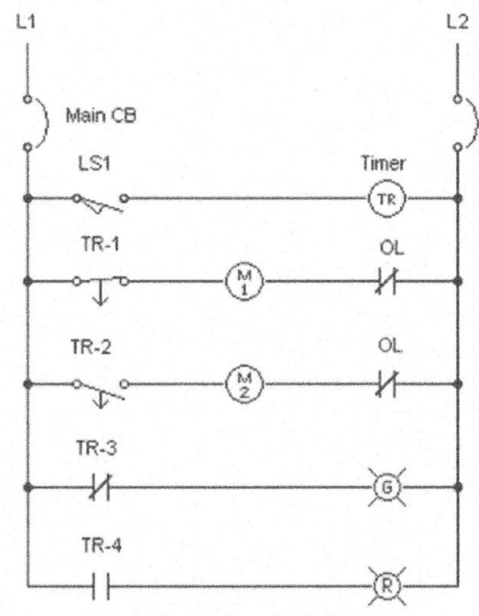

Off-delay timer instantaneous and delayed switching.

16. The purpose of this assignment is to analyze the **operation of the one-shot timer relay** shown. Download the analysis assignment circuit file **CH7 Assign 16** from the **Diagrams-Analysis-NEMA** folder. Click on the Main Switch to turn on the simulation.

   a. Close the circuit breaker (CB).
      Toggle the Proximity Switch ON for 1 second and then OFF.
      The length of time that the solenoid valve stays energized during this operation is _______ seconds.

   b. Reset the process by switching the circuit breaker (CB) OFF and ON.
      Repeat the operation with the Proximity Switch ON for 5 seconds and then OFF.
      The length of time that the solenoid valve stays energized during this operation is _______ seconds.

   c. Reset the process by switching the circuit breaker (CB) OFF and ON.
      Repeat the operation with the Proximity Switch ON for 10 seconds and then OFF.
      The length of time that the solenoid valve stays energized during this operation is _______ seconds.

   d. Reset the process by switching the circuit breaker (CB) OFF and ON.
      Repeat the operation with the Proximity Switch ON for 15 seconds and then OFF.
      The length of time that the solenoid valve stays energized during this operation is _______ seconds.

   e. Describe how this circuit is designed to operate.

      ____

      ____

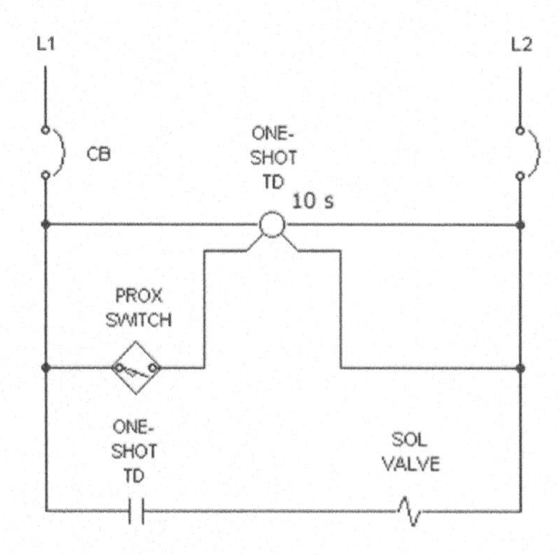

One-shot timer relay.

17. The purpose of this assignment is to analyze the **operation of the recycle timer relay** shown. Download the analysis assignment circuit file **CH7 Assign 17** from the **Diagrams-Analysis-NEMA** folder. Click on the Main Switch to turn on the simulation. Close the circuit breaker (CB).

a. What event triggers the operation of the timer circuit?

____

____

b. What is the ON/OFF flashing rate of the pilot light?

i. ON __________ ii. OFF __________

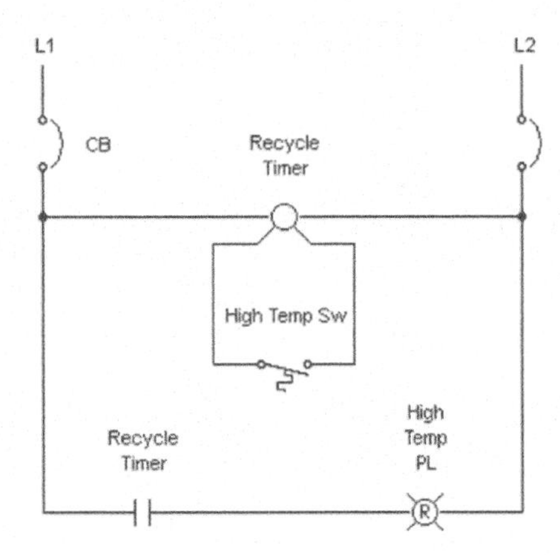

Recycle timer relay.

TROUBLESHOOTING ASSIGNMENTS

1. This assignment involves the **troubleshooting** of a **faulty operative** relay switching circuit. Download the troubleshooting assignment circuit file **CH7 T01** from the **Diagrams-Troubleshooting-NEMA** folder.

   a. Turn on the Main CB. Operate the circuit and describe the faulty condition that exists.

   ____

   ____

   b. What type of fault (open or short) is most likely? Why?

   ____

   ____

   c. Which type of component (switch or load) is most likely at fault? Why?

   ____

   ____

   d. From your understanding of how the circuit is supposed to operate, which two load devices should be suspected? Why?

   ____

   ____

   e. With the Main CB open, use the test probe Continuity mode to test and record which component is at fault.

   ____

   (Turn the Main Power Switch off, replace the defective component, and operate the circuit to verify your answer.)

2. This assignment involves the **troubleshooting** of a **faulty operative** instantaneous and delayed relay switching circuit. Download the troubleshooting assignment circuit file **CH7 T02** from the **Diagrams-Troubleshooting-NEMA** folder.

   a. Turn on the Main CB. Operate the circuit and describe the faulty condition that exists.

      ____

      ____

   b. With the circuit in the timed-out state, use the test probe Power function to record the (ON/OFF) state of the following test points:

      i. L2 to the line side of the TR-3 contact. ____________________

      ii. L2 to the load side of the TR-3 contact. ______________

      iii. Across the Y light. __________________

   c. Based on your readings, what is the most likely problem? Why?

      ____

      ____

      ____

      (Turn the Main Power Switch off, replace the defective component, and operate the circuit to verify your answer.)

3. This assignment involves the **troubleshooting** of a **faulty operative** 10 second on-delay timer circuit. Download the troubleshooting assignment circuit file **CH7 T03** from the **Diagrams-Troubleshooting-NEMA** folder.
   a. Turn on the Main CB. Operate the circuit and describe the faulty condition that exists.

      ____

      ____

   b. With the circuit in the timed-out state, use the test probe Power function to record the (ON/OFF) state of the following test points:
      i. L2 to the load side of the normally open TR1-1 contact. _______
      ii. L2 to the load side of the normally closed TR1-2 contact. ___________
      iii. Across the TR coil. _______________
   c. Based on your readings, what is the most likely problem? Why?

      ____

      ____

      (Turn the Main Power Switch off, replace the defective component, and operate the circuit to verify your answer.)

4. This assignment involves the **troubleshooting** of an **inoperative** 10 second off-delay timer circuit. Download the troubleshooting assignment circuit file **CH7 T04** from the **Diagrams-Troubleshooting-NEMA** folder.
   a. Close the main breaker and use the test probe Power function to record the (ON/OFF) power state of the following test points:
      i. L2 to line side of the switch. ____________
      ii. L2 to the load side of the switch (switch off). __________
      iii. L2 to the load side of the switch (switch on). ____________
      iv. Across TR coil (switch on). ____________
      v. L2 to line side of the TR1-1 contact. ______________
      vi. L2 to load side of the TR1-1 contact (switch on). ___________
      vii. Across R pilot light. ____________
      viii. L2 to line side of the TR1-2 contact. _____________
      ix. L2 to load side of the TR1-2 contact (switch on). ___________
      x. Across G pilot light. _____________
   b. Based on your readings, what is the most likely problem?

      ____

      ____

   c. What specific power readings lead you to conclude this?

      ____

      ____

      (Turn the Main Power Switch off, replace the defective component, and operate the circuit to verify your answer.)

5. This assignment involves the **troubleshooting** of a **faulty operative** recycle timer circuit. Download the troubleshooting assignment circuit file **CH7 T05** from the **Diagrams-Troubleshooting-NEMA** folder.
   a. Turn on the Main CB. Operate the circuit and describe the faulty condition that exists.

      ____

   b. What type of fault condition likely exists?

      ____

   c. What two components should be suspect?

      ____

   d. Conduct a continuity test of each suspected component and record which one is faulted and why.

      ____

      (Turn the Main Power Switch off, replace the defective component, and operate the circuit to verify your answer.)

6. This assignment involves the **troubleshooting** of a **faulty operative** one-shot timer circuit. Download the troubleshooting assignment circuit file **CH7 T06** from the **Diagrams-Troubleshooting-NEMA** folder.
   a. The reported problem is that when the CB is first closed the R light comes on, stays on for a period of time, then turns off. The circuit than appears to be locked in that off state. Operate the circuit to verify this problem. Open the main CB and use the test probe Continuity function to record the (Yes/No) continuity state of the following test points:
      i. Across PB (not activated). ____________
      ii. Across PB (activated). ____________
      iii. Across CR1-1 contact ____________
      iv. Across TR1-1 contact ____________
      v. Across CR1-2 contact ____________
      vi. Across TR2-1 contact ____________
      vii. Across TR2-2 contact ____________
   b. Based on your readings, what is the most likely problem? Why?

      ____

      ____

      (Turn the Main Power Switch off, replace the defective component, and operate the circuit to verify your answer.)

7. This assignment involves the **troubleshooting** of an **inoperative** latching relay circuit. Download the troubleshooting assignment circuit file **CH7 T07** from the **Diagrams-Troubleshooting-NEMA** folder.
   a. Operate the circuit to confirm this problem. With the main CB closed use the test probe Power function to record the (ON/OFF) power state of the following test points:
      i. L2 to the line side of the latch button. ____________
      ii. L2 to the load of the latch button (button not activated). ______
      iii. L2 to the load of the latch button (button activated). ________
      iv. L2 to the line side of the reset button. ____________
      v. L2 to the load side of the reset button (button not activated). _________
      vi. L2 to the load side of the reset button (button activated). ___________
      vii. L2 to the line side of the contact. ______________
      viii. L2 to the load side of the contact. _____________
   b. Based on your readings, what is the most likely problem? Why?

      ____

      ____

      (Turn the Main Power Switch off, replace the defective component, and operate the circuit to verify your answer.)

8. This assignment involves the **troubleshooting** of a **faulty operative** alternating relay circuit. Download the troubleshooting assignment circuit file **CH7 T08** from the **Diagrams-Troubleshooting-NEMA** folder.

   a. Turn on CB and operate the circuit. Describe the faulty condition that exists.

   ____

   ____

   ____

   b. With CB closed use the test probe Power function to record the (ON/OFF) power state of the following test points:

      i. L2 to the L1 side of the coil. __________

      ii. L2 to the line side of the pushbutton. __________

      iii. L2 to the load of the pushbutton (button not activated). __________

      iv. L2 to the load of the pushbutton (button activated). __________

      v. L2 to the T terminal of the coil (button activated). __________

   c. Based on your readings, what is the most likely problem? Why?

   ____

   ____

   ____

   (Turn the Main Power Switch off, replace the defective component, and operate the circuit to verify your answer.)

9. This assignment involves the **troubleshooting** of a **faulty operative** alternating relay motor control circuit. Download the troubleshooting assignment circuit file **CH7 T09** from the **Diagrams-Troubleshooting-NEMA** folder.
   a. Turn on the CB and operate the circuit. Describe the operating condition that exists on the:
      i. First start-up of the process.

      ____

      ii. Second start-up of the process.

      ____

      iii. Third start-up of the process.

      ____

   b. With the main CB closed use the test probe Power function to check the control part of the circuits by recording the (ON/OFF) power state of the following test points:
      i. Across M1 and M2 starter coils on the first start-up.

      M1 ____________ M2 ____________

      ii. Across M1 and M2 starter coils on the second start-up.

      M1 ____________ M2 ____________

      iii. Across M1 and M2 starter coils on the third start-up.

      M1 ____________ M2 ____________

      iv. Based on your readings, what, if any, fault is evident in the control circuit?

      ____

   c. With the main CB closed use the test probe Power function to check the power part of the circuit by recording the (ON/OFF) power state of the following test points:
      i. Across the line side of the three M1contacts on the first start-up.

      ____

      ii. Across the load side of the three M1contacts on the first start-up.

      ____

      iii. Across the line side of the three M2 contacts on the second start-up.

      ____

      iv. Across the load side of the three M2 contacts on the second start-up.

      ____

      v. Based on your readings, what, if any, fault is evident in the power circuit?

      ____

      (Turn the Main Power Switch off, replace the defective component, and operate the circuit to verify your answer.)

10. This assignment involves the **troubleshooting** of a **faulty operative** rungs of a motor control circuit. Download the troubleshooting assignment circuit file **CH7 T10** from the **Diagrams-Troubleshooting-NEMA** folder. Each of the rungs contains a single fault. With the main CB closed use the test probe Power function to check for each rung fault. State the fault and the specific reading that led you used to this conclusion.

    a. Rung A i. Fault: ____________________

    ii. Reason: ____

    b. Rung B i. Fault: ____________________

    ii. Reason: ____

    ____

    c. Rung C i. Fault: ____________________

    ii. Reason: ____

    ____

    d. Rung D i. Fault: ____

    ii. Reason: ____

    ____

    e. Rung E i. Fault: ____________________

    ii. Reason: ____

    ____

    (Turn the Main Power Switch off, replace the defective component, and operate the circuit to verify your answer.)

11. This assignment involves the **troubleshooting** of a **faulty operative** interposing relay current control circuit. Download the troubleshooting assignment circuit file **CH7 T12** from the **Diagrams-Troubleshooting-NEMA** folder.

    a. Close the main breaker and summarize the faulty operating condition that exists.

    ____

    ____

    b. Which section of the circuit (power/control) should be checked first?

    ____

    c. Open the main breaker and use the test probe Continuity function to record the (Yes/NO) continuity state of the following test points:

       i. Across the Temp Switch (switch open). ______________

       ii. Across the Temp Switch (switch closed). ________________

       iii. Across the CR contact. _________________

    d. Base on your readings, what is the problem?

    ____

    e. Explain what causes the motor to run at all times as a result of this fault.

    ____

    ____

    (Turn the Main Power Switch off, replace the defective component, and operate the circuit to verify your answer.)

12. This assignment involves the **troubleshooting** of an **inoperative** interposing relay voltage control circuit. Download the troubleshooting assignment circuit file **CH7 T13** from the **Diagrams-Troubleshooting-NEMA** folder.

    a. With main breaker and temperature switch both closed use the test probe Power function to record the (ON/OFF) power state of the following circuit test points:

        i. L2 to the L1 side of the control circuit. ____________

        ii. L2 to the load side of the CR coil. ____________

        iii. L2 to the line side of the CR contact. ________________

        iv. L2 to the load side of the CR contact. ______________

        v. L2 to the CR contact side of the M coil. _______________

        vi. L1 to the L2 side of the OL contact. _________________

        vii. L1 to the load side of the OL contact. ____________

        viii. L1 to the contact side of the M coil. ______________

    b. Based on your readings, what is the most likely problem?

    ____

    c. What led you to conclude this to be the problem?

    ____

    ____

    (Turn the Main Power Switch off, replace the defective component, and operate the circuit to verify your answer.)

13. This assignment involves the **troubleshooting** of a **faulty operative** on-delay timer-warning signal circuit. Download the troubleshooting assignment circuit file **CH7 T14** from the **Diagrams-Troubleshooting-NEMA** folder.
    a. Close the main breaker and switch. Operate the circuit and summarize the faulty operating condition that exists.

    ____

    ____

    b. With the circuit still operating, use the test probe Power function to record the (ON/OFF) power state of the following circuit test points:
        i. Across CR control relay. ____________
        ii. Across TR timer relay. ____________
        iii. From the line side of the M starter coil to the line side of the timer relay contact. ______________
        iv. From the line side of the M starter coil to the coil side of the timer relay contact. ________________
    c. Which reading is an indication of a problem? Why?

    ____

    ____

    ____

    d. Which component is at fault? ___________________________________

    (Turn the Main Power Switch off, replace the defective component, and operate the circuit to verify your answer.)

14. This assignment involves the **troubleshooting** of a **faulty operative** timed and instantaneous control circuit. Download the troubleshooting assignment circuit file **CH7 T15** from the **Diagrams-Troubleshooting-NEMA** folder.

    a. Close the main breaker and operate the circuit. Based on your understanding of how the circuit is supposed to operate, describe what initial operating problem occurs.

    ____

    ____

    b. What is the most obvious component to suspect?

    ____

    c. Open the main breaker and use the test probe Continuity function reading to describe the defective state of the component.

    ____

    ____

    (Turn the Main Power Switch off, replace the defective component, and operate the circuit to verify your answer.)

15. This assignment involves the **troubleshooting** of a **faulty operative** one-shot timer relay circuit. Download the troubleshooting assignment circuit file **CH7 T16** from the **Diagrams-Troubleshooting-NEMA** folder.

    a. Turn on the main and circuit CB. and operate the circuit. Close the proximity switch. What happens?

    ____

    b. What is the most likely component fault?

    ____

    c. With the circuit still in this state, use the test probe Continuity function reading to identify the faulted component.

    ____

    (Turn the Main Power Switch off, replace the defective component, and operate the circuit to verify your answer.)

16. This assignment involves the **troubleshooting** of a **faulty operative** recycle timer relay circuit. Download the troubleshooting assignment circuit file **CH7 T17** from the **Diagrams-Troubleshooting-NEMA** folder.
    a. Turn on the main and circuit CB. Operate the circuit the fault condition that exists.

    ____

    ____

    b. With the circuit in this state, use the test probe Power function to record the (ON /OFF) power state of the following test points:
        i. Across the recycle timer coil. ____________
        ii. Across the high temperature switch. ____________
        iii. From L2 to the coil side of the recycle timer contact. ____________
    c. Based on your readings, which of the three readings is suspect. Why?

    ____

    ____

    d. Replace the suspected component and repeat the power test of it. What is the power response to this replacement.

    ____

8 Motor Control Circuits

CIRCUIT ANALYSIS ASSIGNMENTS

1. The purpose of this assignment is to analyze the **operation of the magnetic across-the-line starter**, wired for two start/stop stations shown. Download the analysis assignment circuit file **CH8 Assign 01** from the **Diagrams-Analysis-NEMA** folder. Click on the Main Power Switch (ON/OFF) located on the left side of the toolbar. Left-click on the circuit breaker to close it and on the motor start and stop push buttons to operate them.
   a. How are the two start push buttons connected relative to each other?

      ____

   b. How are the two stop push buttons connected relative to each other?

      ____

   c. Right-click on one of the stop buttons and lock it in the open position. What happens when the circuit is operated? Why?

      ____

      ____

   d. Connect the ground fault to the coil side of the start push button. What happens when the circuit is operated? Why?

      ____

      ____

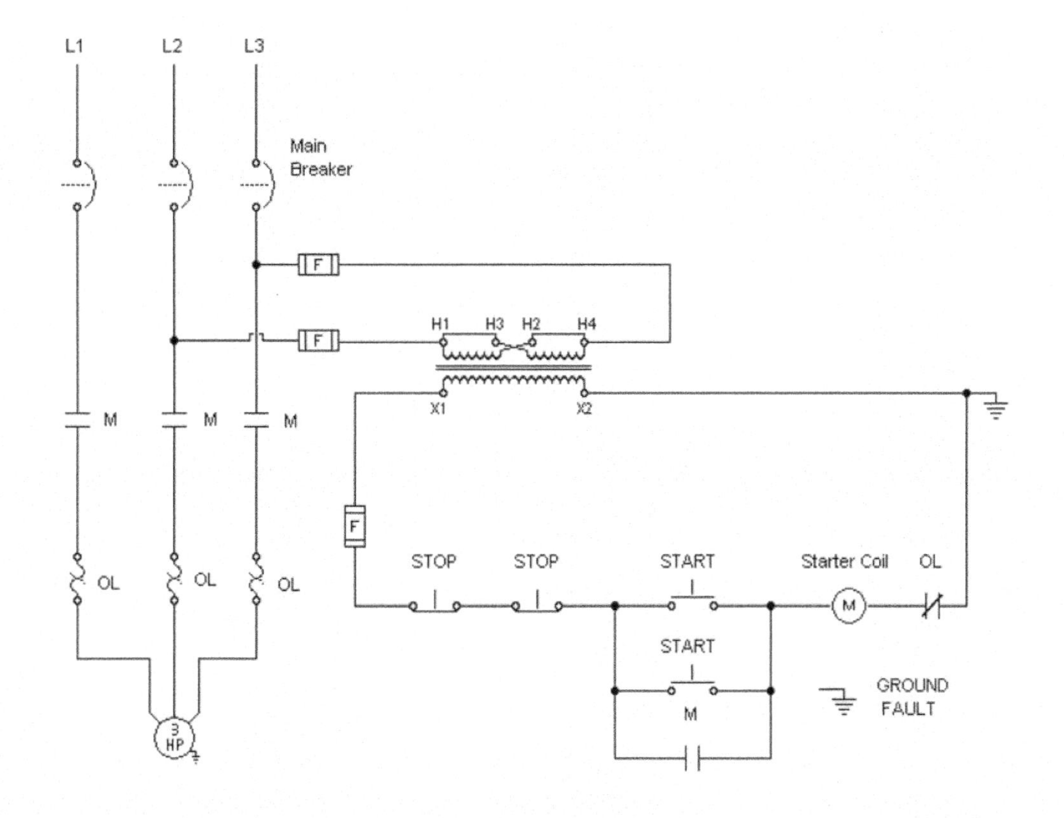

Across-the-line starter circuit with two start/stop stations.

2. The purpose of this assignment is to analyze the **automatic sequential starting of two motors at full line voltage** shown. In order to reduce the amount of starting current, the circuit has been designed so that there will be a short time-delay period between the starting of motor 1 and motor 2. Download the analysis assignment circuit file **CH8 Assign 02** from the **Diagrams-Analysis-NEMA** folder. Click on the Main Power Switch (ON/OFF) located on the left side of the toolbar. Left-click on the circuit breaker to close it and on the motor start and stop button to operate them.
   a. What is time-delay period between the starting of the two motors?

      ____

b. With both motors operating, left-click on motor OL-2 to lock it open. Describe how this affects the operation of the circuit while running and on restarting.

____

____

c. With both motors operating, left-click on motor OL-1 to lock it open. Describe how this affects the operation of the circuit while running and on restarting. ____

____

d. Left-click on the TR coil to simulate an open in the coil. What happens? Why? ____

____

e. Right-click on starter coil M1 to simulate a shorted coil fault. What happens? Why? ____

____

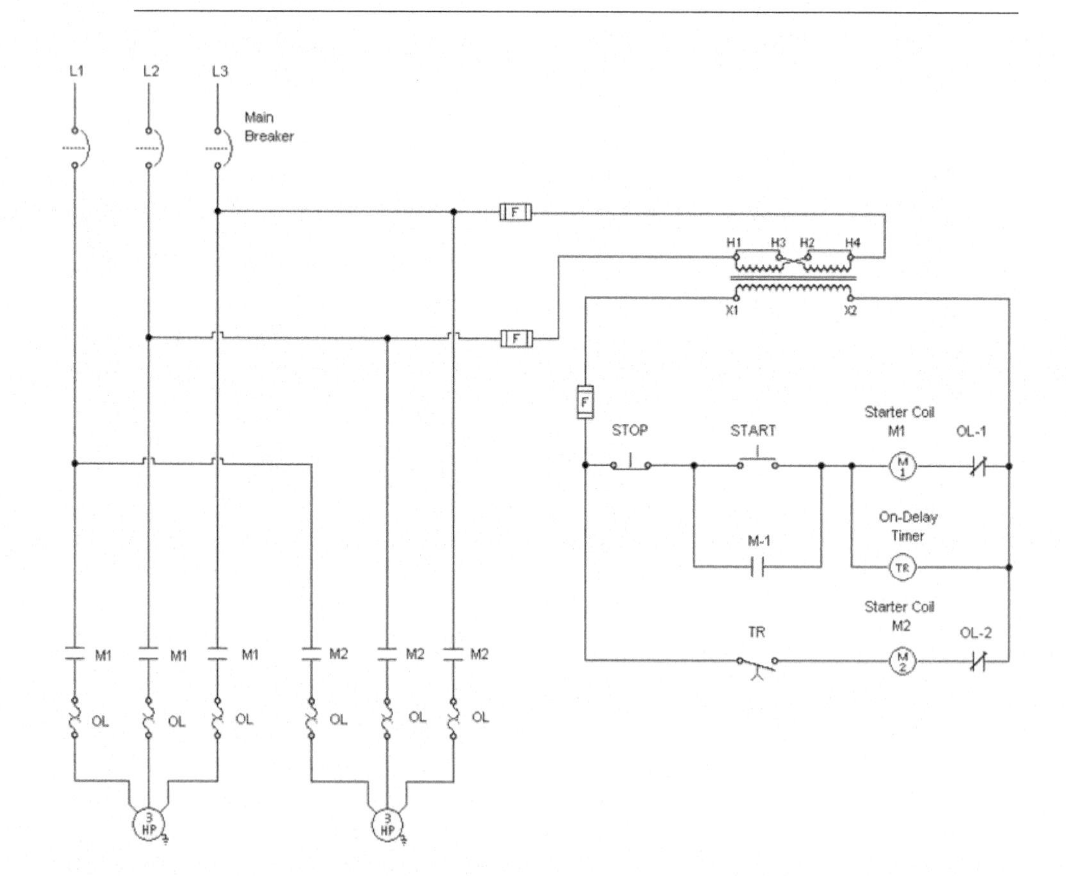

Automatic sequential starting of two motors.

3. The purpose of this assignment is to analyze the **automatic sequential stopping of two motors** shown. Download the analysis assignment circuit file **CH8 Assign 03** from the **Diagrams-Analysis-NEMA** folder. Click on the Main Power Switch (ON/OFF) located on the left side of the toolbar. Left-click on the circuit breaker to close it and on the motor start and stop buttons to operate them.

   a. Run the circuit simulation and summarize the sequence of operations.

   ____

   ____

   ____

   ____

   b. Right-click the normally closed M-1 contact to lock it in the closed position. In what way does this change the operation of the circuit?

   ____

   ____

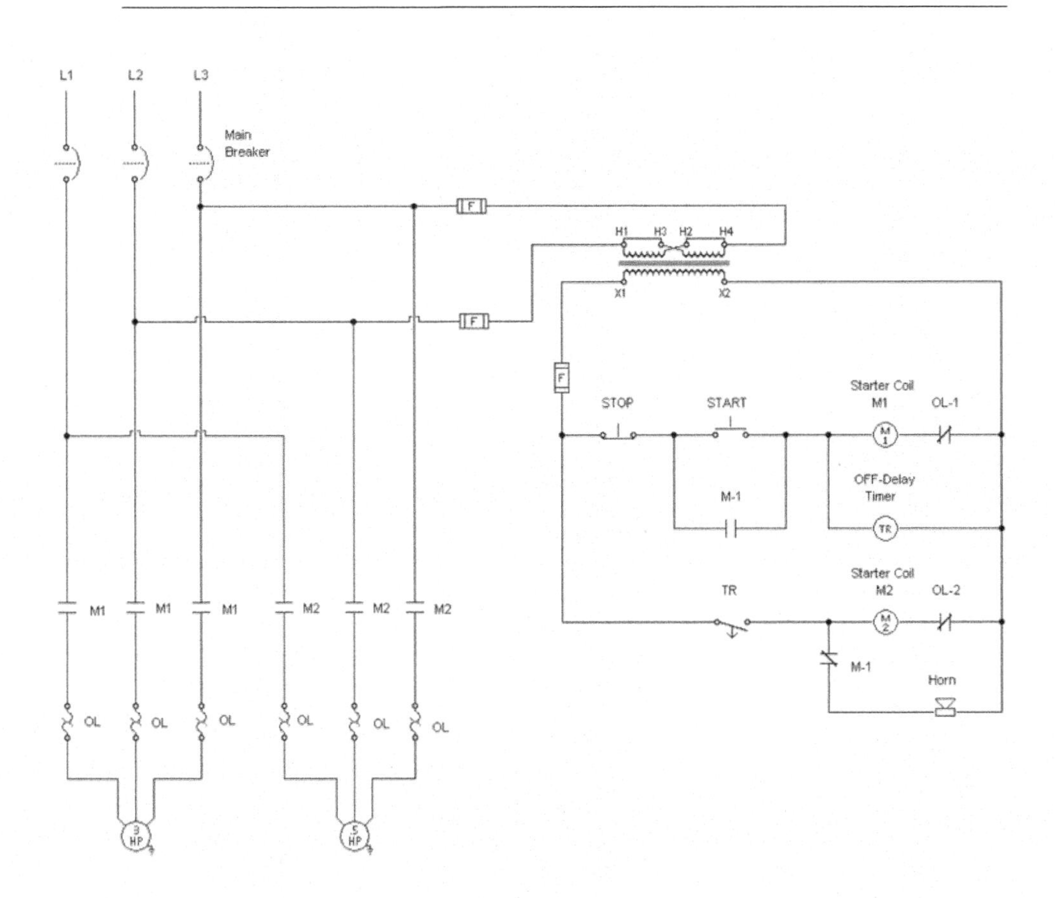

Automatic sequential stopping of two motors.

4. The purpose of this assignment is to analyze the **operation of the primary resistance AC starter circuit** shown. Download the analysis assignment circuit file **CH8 Assign 04** from the **Diagrams-Analysis-NEMA** folder. Click on the Main Power Switch (ON/OFF) located on the left side of the toolbar. Left-click on the circuit breaker to close it and on the motor start and stop buttons to operate them.

   a. What is the preset value of the time-delay period?

   ____

   b. What occurs in the power circuit when the run contactor coil is energized?

   ____

   ____

   c. Left-click on the starting resistors to simulate an open fault in each of them. In what way does this change the operation of the circuit?

   ____

   ____

   ____

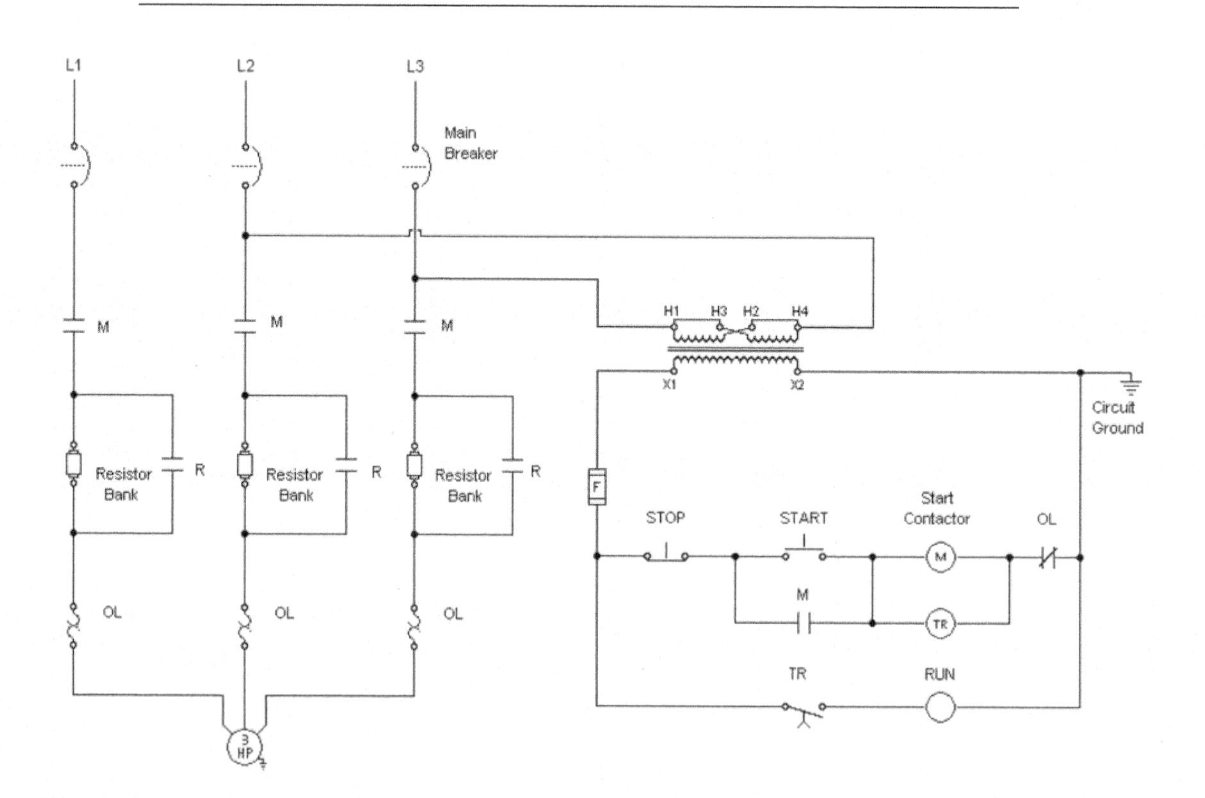

Primary resistance AC starter circuit.

5. The purpose of this assignment is to analyze the **operation of the wye-delta starter circuit** shown. Download the analysis assignment circuit file **CH8 Assign 05** from the **Diagrams-Analysis-NEMA** folder. Click on the Main Power Switch (ON/OFF) located on the left side of the toolbar. Left-click on the circuit breaker to close it and on the motor start and stop buttons to operate them.
   a. Which coil(s) is/are energized when the motor is first started?

      ____

   b. Which coil(s) is/are energized after the time-delay period has elapsed?

      ____

   c. Left-click on relay coil TR to fault it open. In what way does this change the operation of the motor?

      ____

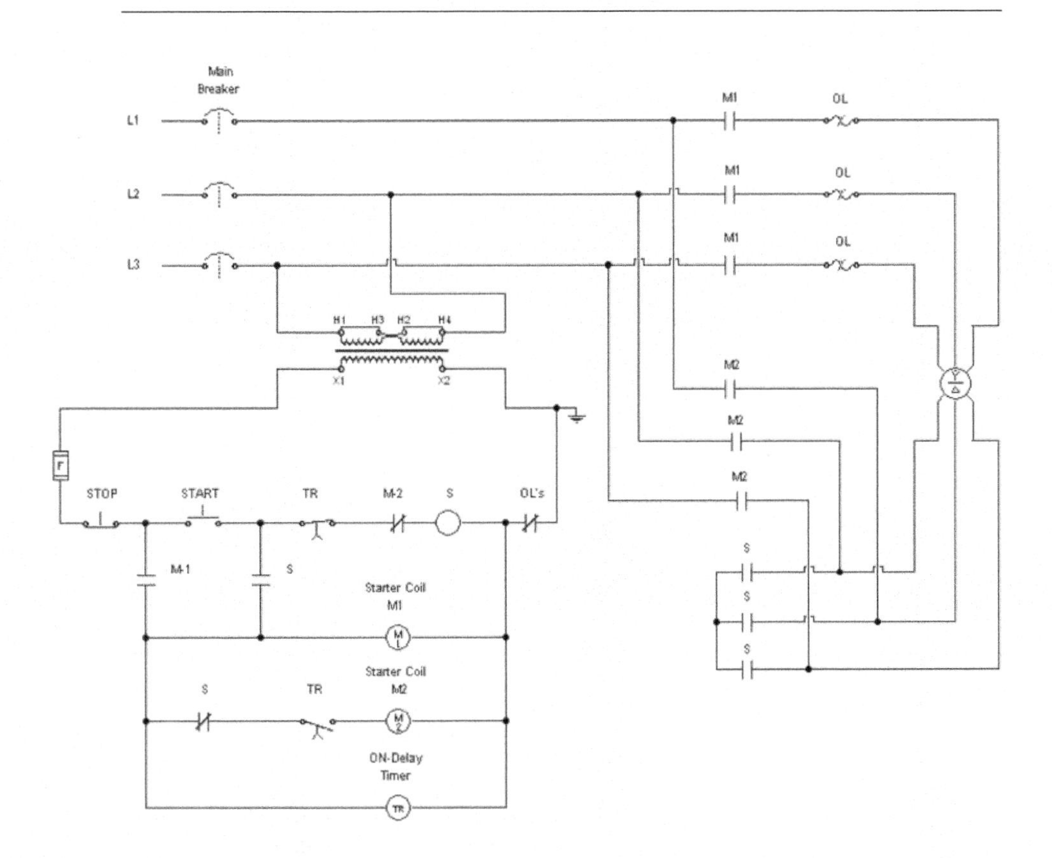

Wye-delta starter circuit.

6. The purpose of this assignment is to analyze the **operation of the reversing starter with auxiliary contact interlocking** shown. Download the analysis assignment circuit file **CH8 Assign 06** from the **Diagrams-Analysis-NEMA** folder. Click on the Main Power Switch (ON/OFF) located on the left side of the toolbar. Left-click on the circuit breaker to close it and on the motor forward and reverse buttons to operate them.

   a. What prevents the reverse contactor coil from being energized when the motor is operating in the forward direction?

      ____

      ____

   b. What prevents the forward contactor coil from being energized when the motor is operating in the reverse direction?

      ____

      ____

   c. With the motor operating in either direction, how is the direction of rotation reversed?

      ____

      ____

   d. Click on the forward NC contactor auxiliary contact and lock it in the open state. In what way is the operation of the circuit altered?

      ____

      ____

   e. Click on the reverse N.O. contactor auxiliary contact and lock it in the closed state. In what way is the operation of the circuit altered?

      ____

      ____

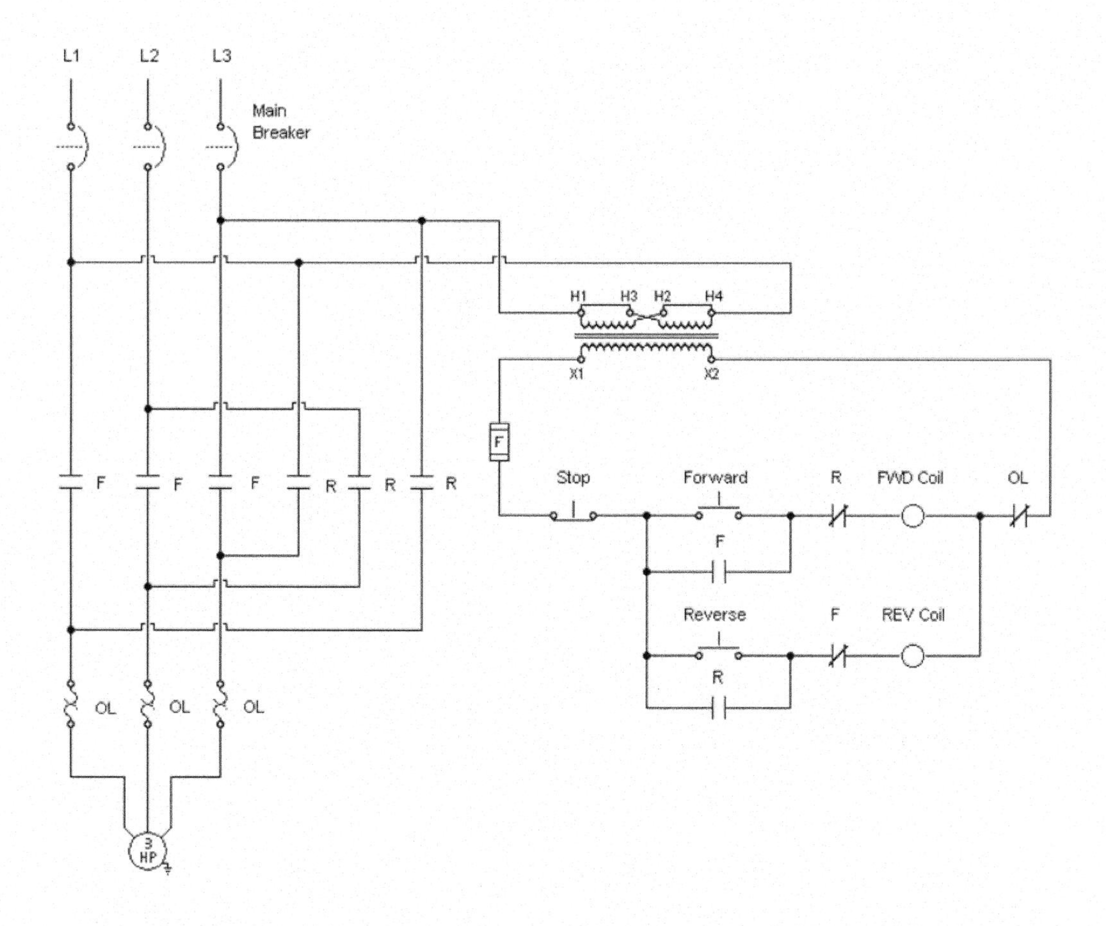

Reversing starter with auxiliary contactor interlocking.

7. The purpose of this assignment is to analyze the **operation of the reversing starter with both electrical auxiliary contact and push button interlocking** shown. Download the analysis assignment circuit file **CH8 Assign 07** from the **Diagrams-Analysis-NEMA** folder. Click on the Main Power Switch (ON/OFF) located on the left side of the toolbar. Left-click on the circuit breaker to close it and on the motor forward and reverse buttons to operate them.
   a. Click on the forward push button. What change of state of the button's contacts occurs?

   ____

   ____

   b. Click on the reverse push button. What change of state of the button's contacts occurs?

   ____

   ____

   c. With the motor operating in the forward direction, try reversing the direction of rotation without first pressing the stop push button. What happens? Why?

   ____

   ____

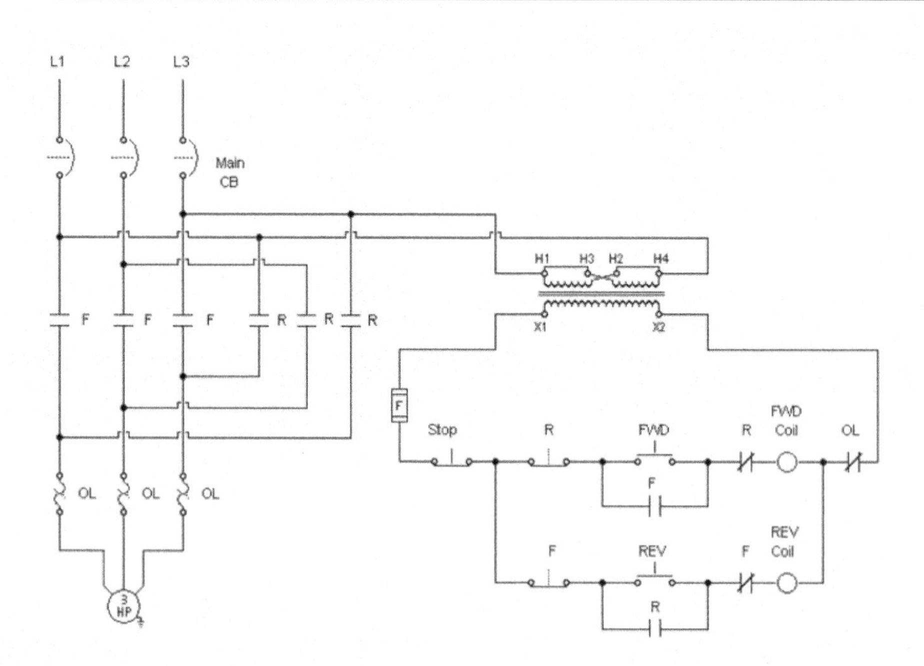

Reversing starter with both auxiliary contact and pushbutton interlocking.

8. The purpose of this assignment is to analyze the **operation of the reversing starter with limit switches connected to limit travel** shown. Download the analysis assignment circuit file **CH8 Assign 08** from the **Diagrams-Analysis-NEMA** folder. Click on the Main Power Switch (ON/OFF) located on the left side of the toolbar. Left-click on the circuit breaker to close it and on the directional buttons and limit switches to operate them.
   a. Operate the circuit with the contacts of both limit switches closed. Why is travel not restricted in either direction?

      ____

   b. Operate the circuit with the contact of the forward limit switch open and the reverse limit switch contact closed. In what way is travel restricted?

      ____

   c. Operate the circuit with the contact of the reverse limit switch open and the forward limit switch contact closed. In what way is travel restricted?

      ____

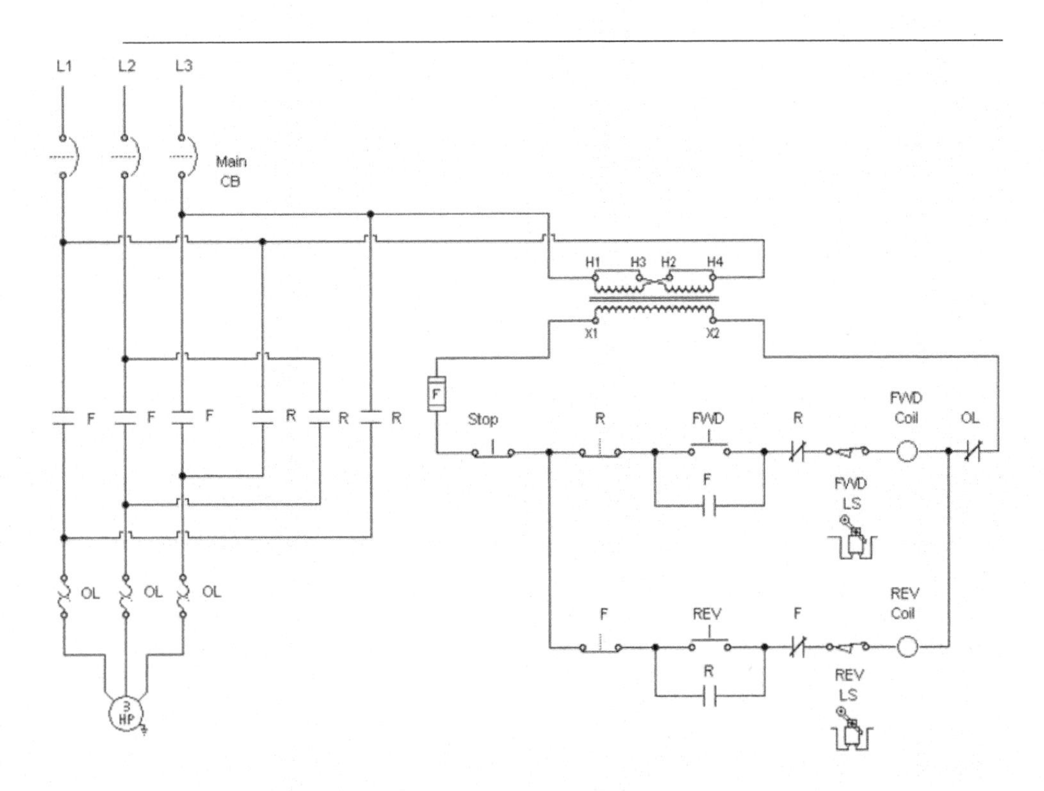

Reversing starter wired with limit switches connected to limit travel.

9. The purpose of this assignment is to analyze the **operation** of the **reversing starter for a reciprocating machine process** shown. Download the analysis assignment circuit file **CH8 Assign 09** from the **Diagrams-Analysis-NEMA**

folder. Click on the Main Power Switch (ON/OFF) located on the left side of the toolbar. Left-click on the circuit breaker to close it and on the buttons and limit switches to operate them.

a. With the motor traveling in the forward direction, what happens when the forward limit switch is actuated? ____________________

b. With the motor traveling in the reverse direction, what happens when the reverse limit switch is actuated? ____________________

c. Assume that the forward limit switch sticks and fails to return to its normal deactivated position once the motor starts operating in the reverse direction. Simulate this scenario. In what way is the operation of the circuit altered as a result of this fault condition?

____

____

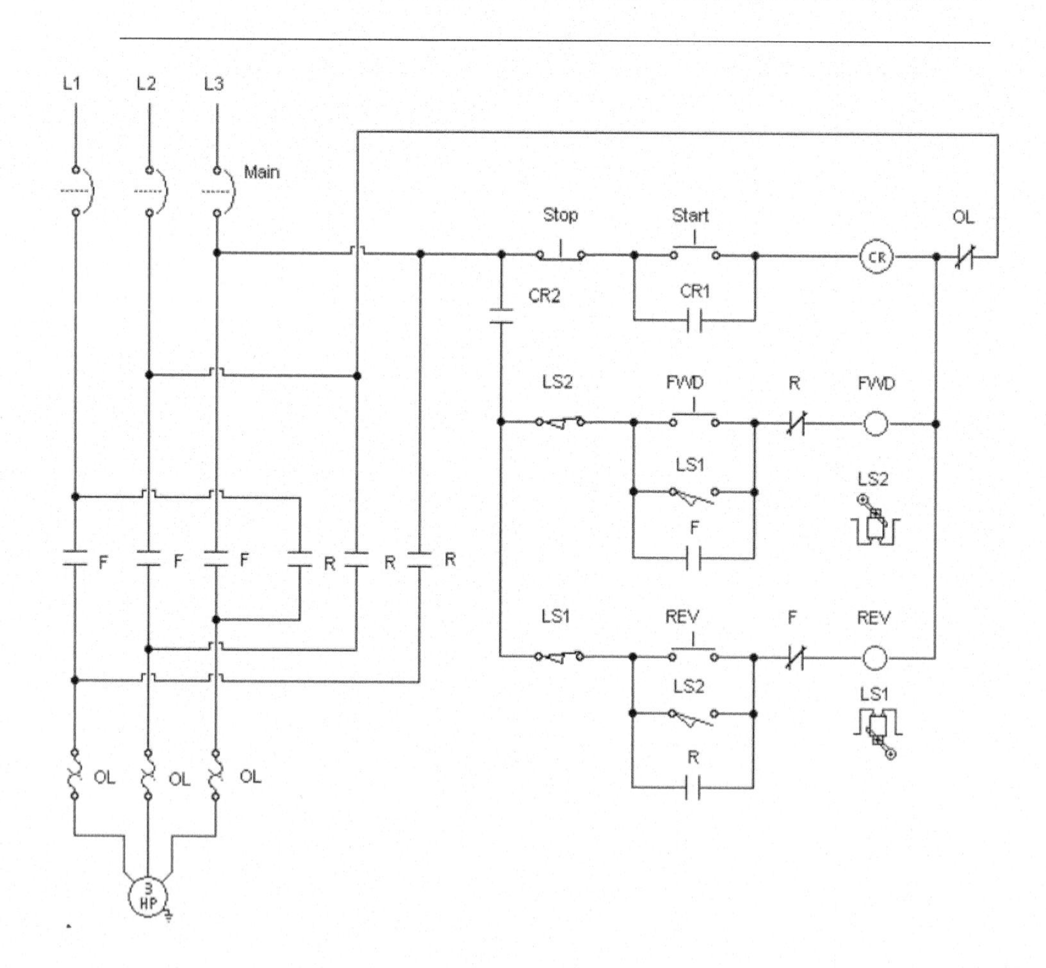

Reciprocating machine process.

10. The purpose of this assignment is to analyze the **operation of the pushbutton jog circuit** shown. Download the analysis assignment circuit file **CH8 Assign 10** from the **Diagrams-Analysis-NEMA** folder. Click on the Main Power Switch (ON/OFF) located on the left side of the toolbar. Left-click on the circuit breaker to close it and on the buttons to operate them.

    a. With the motor operating in the run mode, what happens when the jog push button is actuated? ______________________________

    b. Assume the N.O. make jog contact becomes locked in the open state. Simulate this scenario. In what way is the operation of the circuit altered as a result of this fault condition?

    ____

    ____

    c. Assume that on release of the jog push button, the NC pushbutton contacts reclose before the starter maintaining contact M opens. What potential hazard operating condition would be created?

    ____

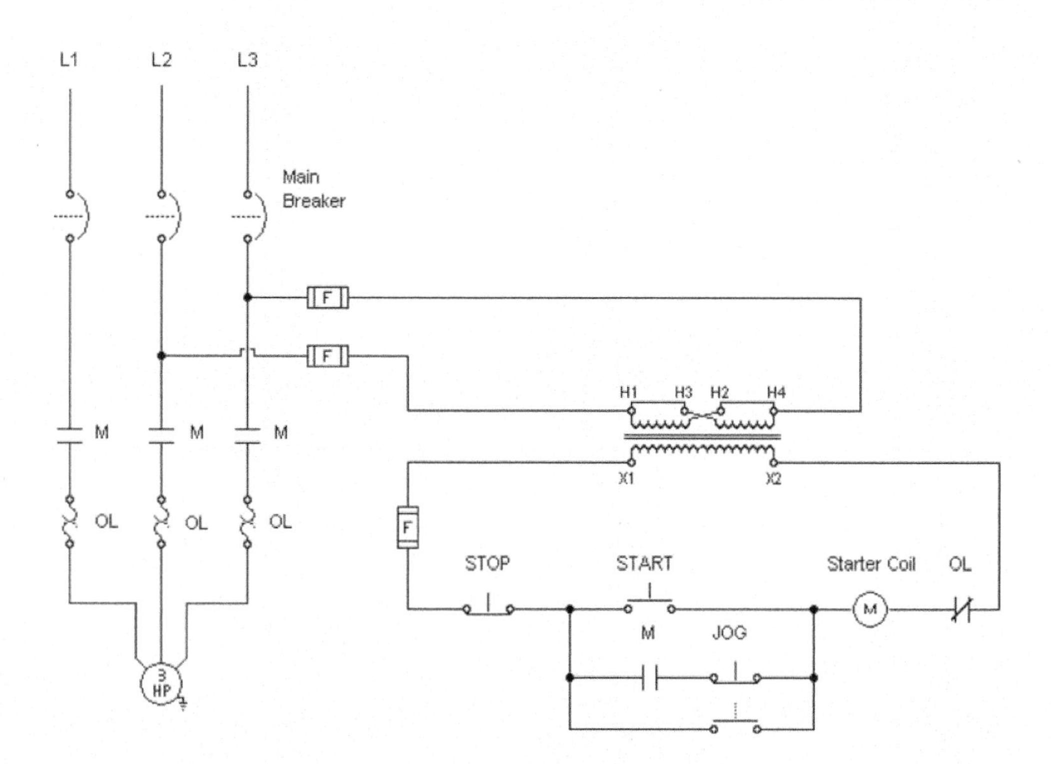

Pushbutton jog circuit.

11. The purpose of this assignment is to analyze the **operation of the control relay jog circuit** shown. Download the analysis assignment circuit file **CH8 Assign 11** from the **Diagrams-Analysis-NEMA** folder. Click on the Main Power Switch (ON/OFF) located on the left side of the toolbar. Left-click on the circuit breaker to close it and on the buttons to operate them.

    a. Which coil is directly energized by closing the start push button?

    ____

    b. With the motor operating in the run mode, what happens when the jog button is pressed? Why?

    ____

    ____

    c. Assume the CR coil is faulted open. Simulate this scenario. In what way is the operation of the circuit altered as a result of this fault condition?

    ____

    d. Assume the M coil is faulted open. Simulate this scenario. In what way is the operation of the circuit altered as a result of this fault condition?

    ____

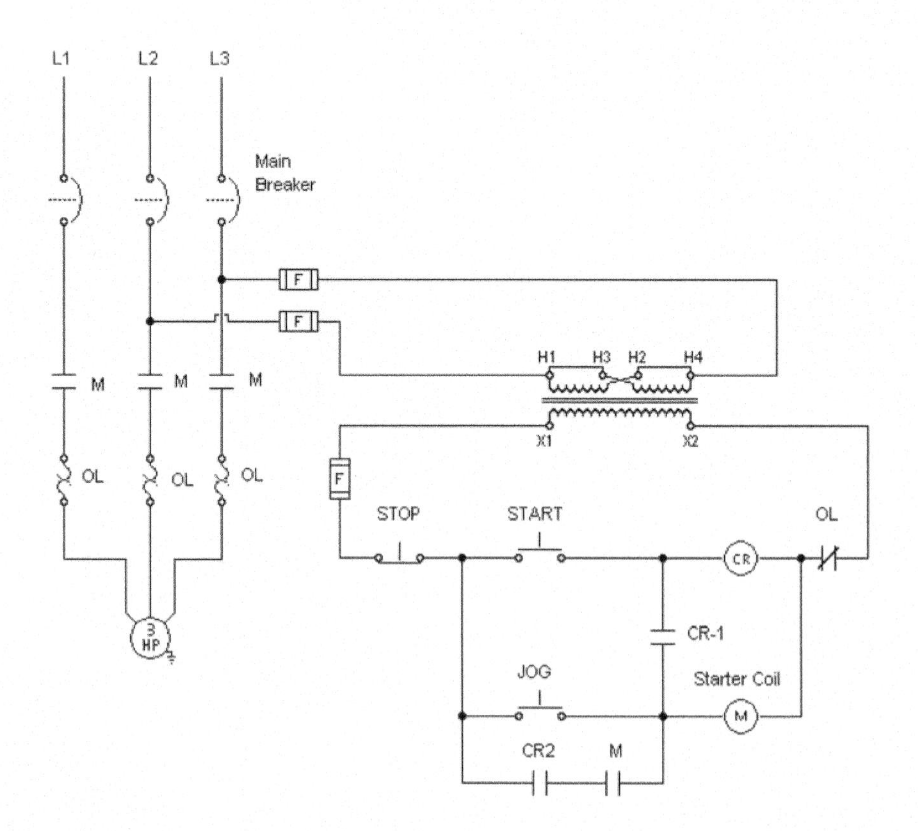

Control relay jog circuit.

12. The purpose of this assignment is to analyze the **operation of the start/stop/selector jog circuit** shown. Download the analysis assignment circuit file **CH8 Assign 12** from the **Diagrams-Analysis-NEMA** folder. Click on the Main Power Switch (ON/OFF) located on the left side of the toolbar. Left-click on the circuit breaker to close it and on the buttons to operate them.

    a. What state must the run/jog switch be in to operate the motor in the jog mode? ______________________________

    b. With the motor operating in the run mode, what happens when the run/jog switch is opened? Why? ______________________________

    c. Assume the stop push button is locked in the open position. Simulate this scenario. In what way is the operation of the circuit altered as a result of this fault condition?

    ____

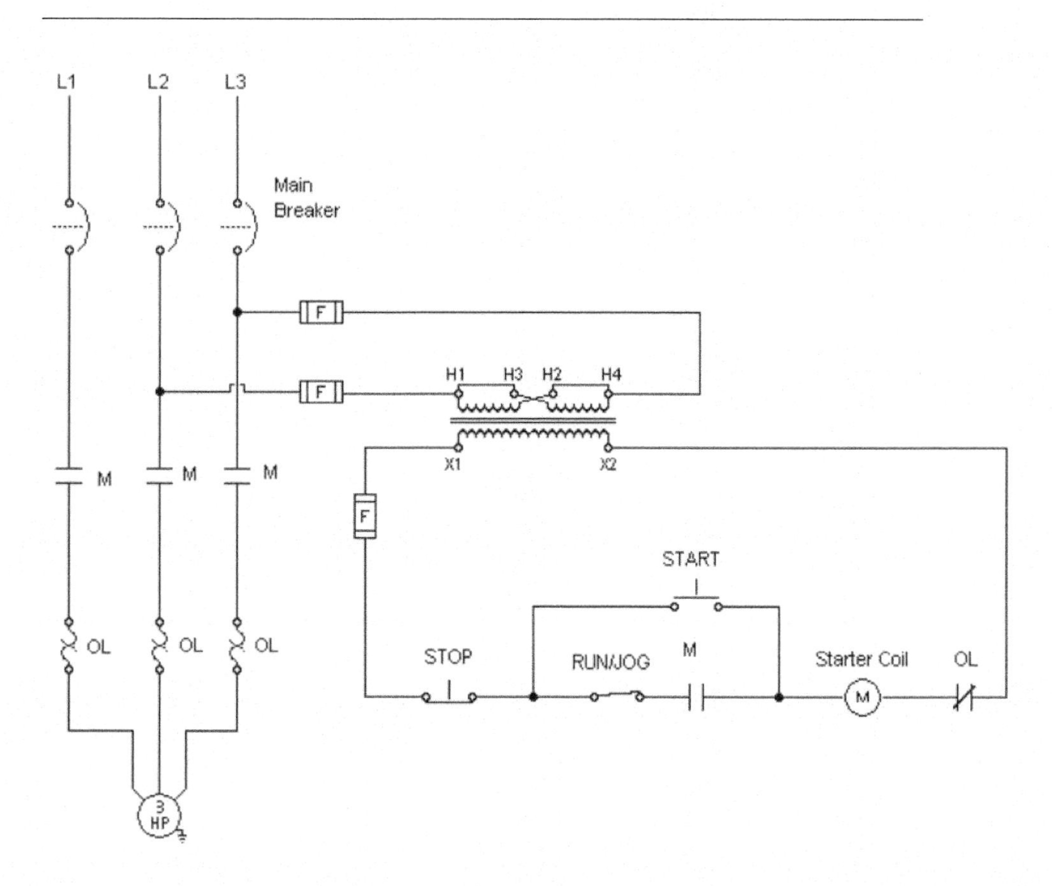

Start/stop/selector jog circuit.

13. The purpose of this assignment is to analyze the **operation of the antiplugging protection circuit** shown. Download the analysis assignment circuit file **CH8 Assign 13** from the **Diagrams-Analysis-NEMA** folder. Click on the Main Power Switch (ON/OFF) located on the left side of the toolbar. Left-click on the circuit breaker to close it and on the switch buttons to operate them. Activate the F and R zero-speed switch contacts with respect to the speed and direction rotation of the motor.
    a. Immediately after the forward push button is pressed, with the motor operating, what state (open or closed) should the F and R zero-switch contacts be in? Simulate this scenario.

    ____

    b. With the motor running in the forward direction what state should the F and R zero-switch contacts be in immediately after pressing the stop push button? Simulate this scenario.

    ____

    c. At what point after this will pressing the reverse push button allow the reverse contactor to be energized? Simulate this scenario.

    ____

    ____

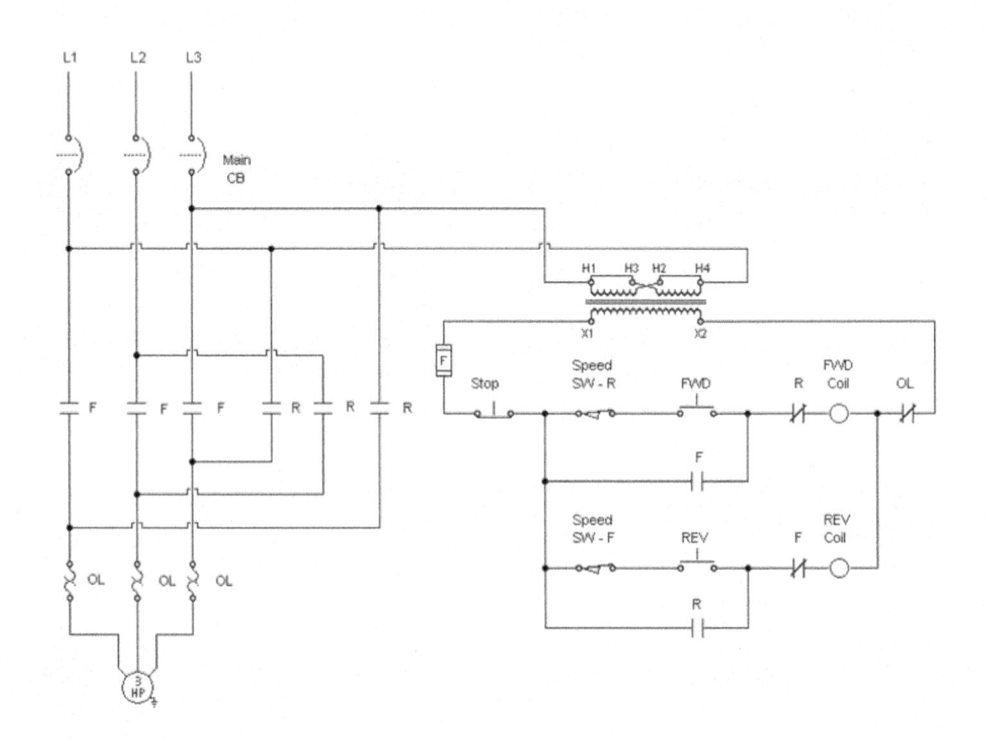

Antiplugging protection circuit.

14. The purpose of this assignment is to analyze the **operation of the dynamic braking circuit for a DC motor** shown. Download the analysis assignment circuit file **CH8 Assign 14** from the **Diagrams-Analysis-NEMA** folder. Click on the Main Power Switch (ON/OFF) located on the left side of the toolbar. Left-click on the circuit breaker to close it and on the start and stop buttons to operate them.

    a. With coil M deenergized, how is the dynamic braking resistor connected relative to the motor armature? ___________________________________

    b. When is the only time that current flows through the dynamic braking resistor?

    ____

    c. Assume the dynamic braking resistor is faulted shorted. How would this affect the period of time for the motor to stop? Why?

    ____

    ____

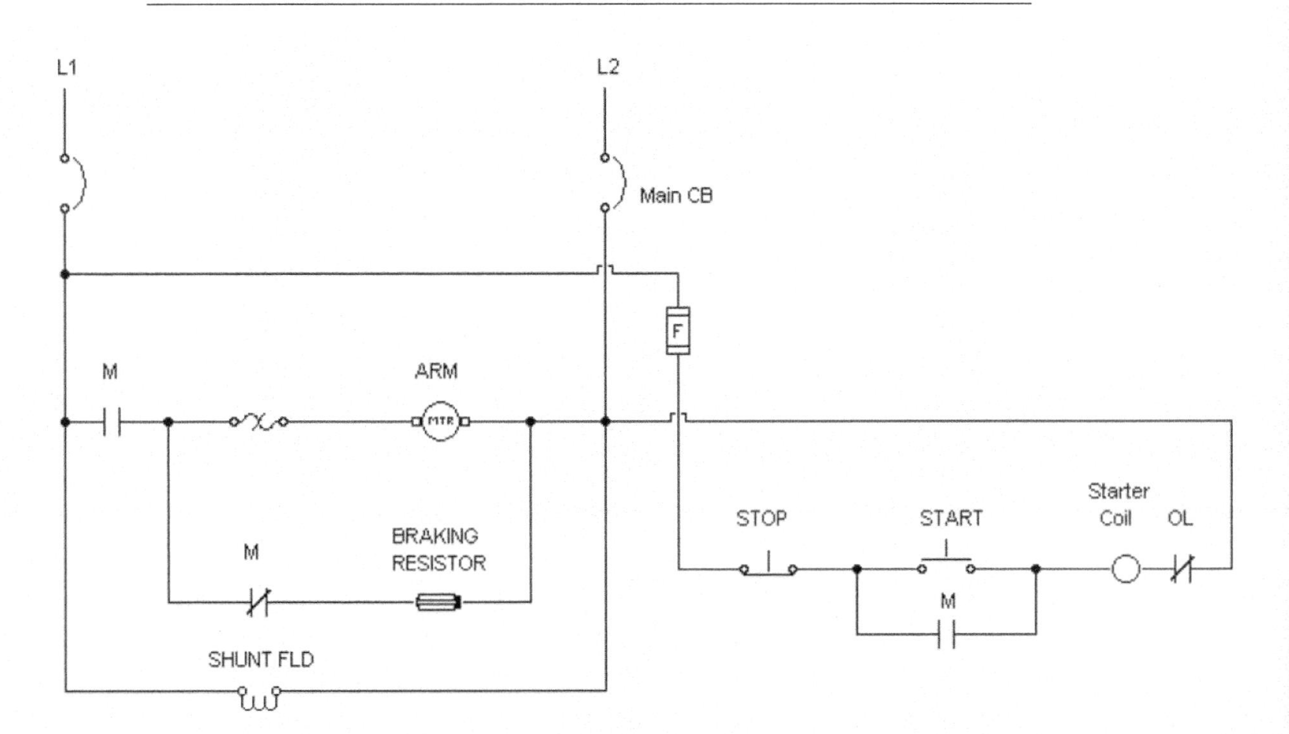

Dynamic braking circuit for a DC motor.

15. The purpose of this assignment is to analyze the **operation of the AC motor DC injection braking circuit** shown. Download the analysis assignment circuit file **CH8 Assign 15** from the **Diagrams-Analysis-NEMA** folder. Click on the Main Power Switch (ON/OFF) located on the left side of the toolbar. Left-click on the circuit breaker to close it and on the start and stop buttons to operate them.
    a. What is the function of the rectifier? ______________________________
    b. What type of pushbutton contact arrangement is used for the start button?

    ____
    c. What is the state of each of the coils immediately after the start push button has been pressed? ______________________________
    d. What is the state of each of the coils immediately after the stop push button has been pressed? ______________________________
    e. What is the state of each of the coils after the injection braking has been applied?

    ____
    f. Assume that one lead on the secondary of the braking transformer is moved to the center tap terminal. Would this increase or decrease the amount of time required to bring the motor to a stop? Why?

    ____

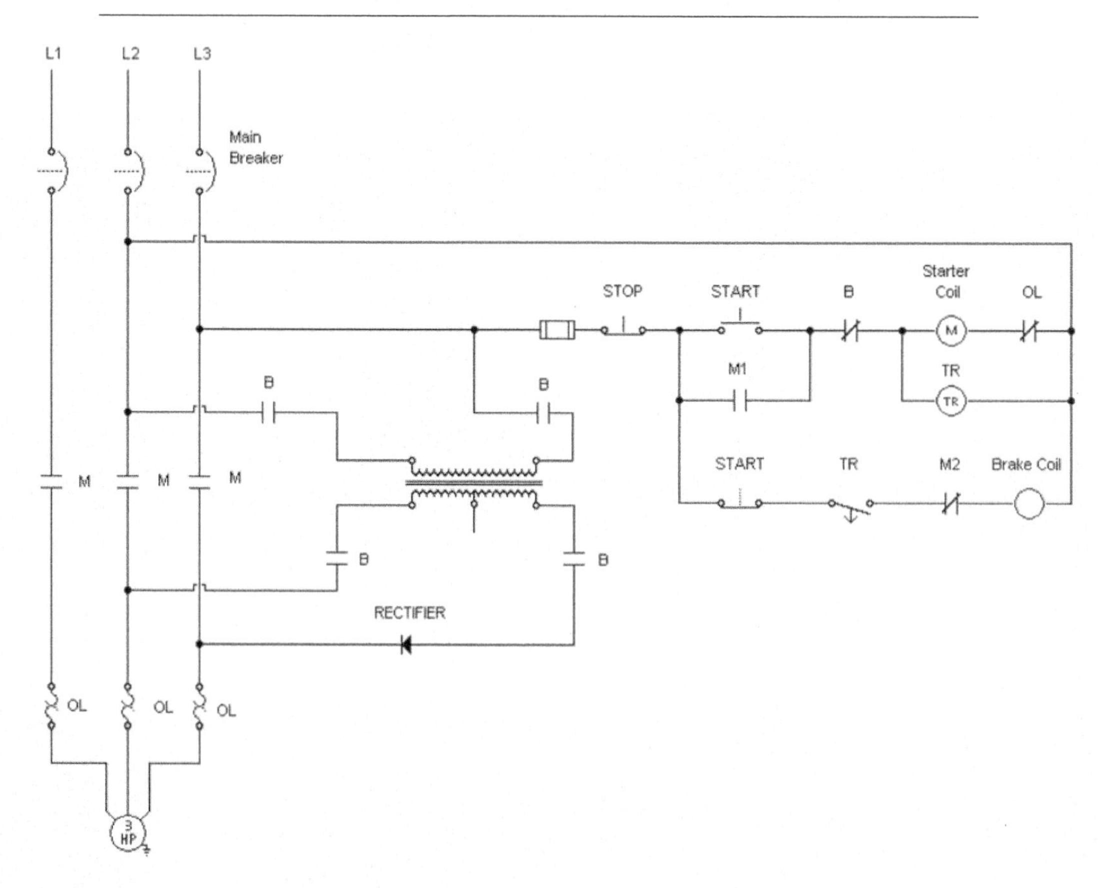

AC motor DC injection braking circuit.

16. The purpose of this assignment is to analyze the **operation of the control circuit interlocked so that the three motors must be turned on in order** shown. Download the analysis assignment circuit file **CH8 Assign 16** from the **Diagrams-Analysis-NEMA** folder. Click on the Main switch to turn on the simulation.

    a. Turn the motors on with a **2-1-3** sequence and record the resultant state (Energized or Deenergized) of each of the following:

       i. M1 ____________ ii. M2 _______________ iii. M3 _____________

    b. Reset and repeat for a **3-2-1** sequence:

       i. M1 ____________ ii. M2 _______________ iii. M3 _____________

    c. Reset and repeat for a **1-2-3** sequence:

       i. M1 ____________ ii. M2 _______________ iii. M3 _____________

    d. With all three motors operating, turn off M2 and record the resultant state of:

       i. M1 ____________ ii. M3 _______________

    e. With all three motors operating, turn off M3 and record the resultant state of:

       i. M1 ____________ ii. M2 _______________

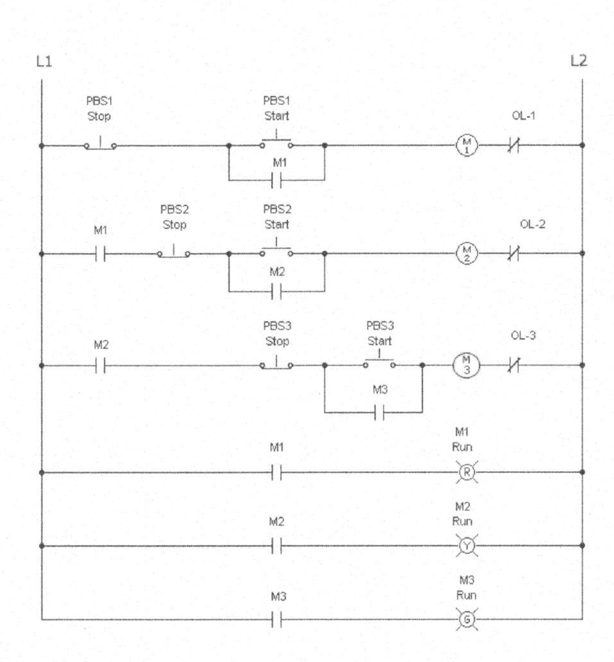

Control circuit interlocked so that the three motors must be turned on in order.

17. The purpose of this assignment is to analyze the **operation of the control circuit interlocked so that the two motors be turned off in order** shown. Download the analysis assignment circuit file **CH8 Assign 17** from the **Diagrams-Analysis-NEMA** folder. Click on the Main switch to turn on the simulation.

    a. Momentarily close the start button and record the state (ON/OFF, Open/Closed, Energized/Deenergized) of each of the following:

       i. M1 coil __________ ii. M2 coil __________ iii. TR coil __________

       iv. M1 (NO) contact __________ v. M1 (NC) contact __________

       vi. TR contact __________ vii. PL ______________

    b. With both motors operating, momentarily open the stop button and record the immediate state of each of the following:

       i. M1 coil __________ ii. M2 coil __________ iii. TR coil __________

       iv. M1 (NO) contact __________ v. M1 (NC) contact __________

       vi. TR contact __________ vii. PL ______________

    c. After the timer has timed off record the state of each of the following:

       i. M1 coil __________ ii. M2 coil __________ iii. TR coil __________

       iv. M1 (NO) contact __________ v. M1 (NC) contact __________

       vi. TR contact __________ vii. PL ______________

    d. What is the time of the off-delay period? __________ seconds.

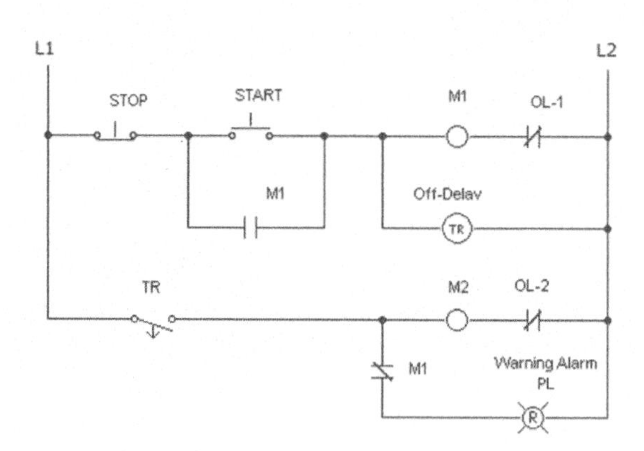

Control circuit interlocked so that the two motors be turned off in order.

18. The purpose of this assignment is to analyze the **operation of the control circuit for Jog/Run selector switch reversing starter** shown. Download the analysis assignment circuit file **CH8 Assign 18** from the **Diagrams-Analysis-NEMA** folder. Click on the main CB to turn on the simulation.

    a. With the selector switch in the **Jog position**, momentarily operate the FWD pushbutton alternately in the closed and open positions. Summarize how the circuit reacts.

    ---

    b. With the selector switch in the **Jog position**, momentarily operate the REV pushbutton alternately in the closed and open positions. Summarize how the circuit reacts.

    ---

    c. With the selector switch in the **Run position**, momentarily operate the FWD pushbutton alternately in the closed and open positions. Summarize how the circuit reacts.

    ---

    d. With the selector switch in the **Run position**, momentarily operate the REV pushbutton alternately in the closed and open positions. Summarize how the circuit reacts.

    ---

    e. What two types of electrical interlocking are used to prevent both the forward and reverse contactors from both being energized at the same time?

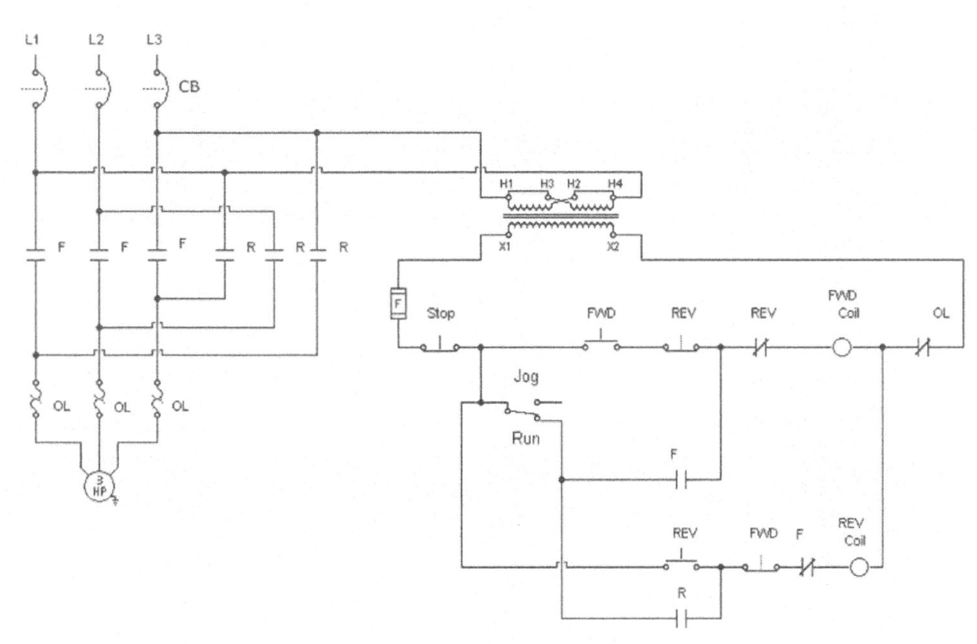

Jog/Run selector switch reversing starter.

19. The purpose of this assignment is to analyze the **operation of the control circuit for anti-plugging executed using time-delay relays** shown. Download the analysis assignment circuit file **CH8 Assign 19** from the **Diagrams-Analysis-NEMA** folder. Click on the main CB to turn on the simulation.

   a. Start the motor in the forward direction and record the state (Open/Closed, Energized/Deenergized) of each of the following:

      i. FWD coil ________ ii. REV coil ________ iii. TD1 coil ________

      iv. TD1 contact ________ v. TD2 coil ________ vi. TD2 contact ________

   b. Stop the motor in the forward direction and record the initial state (Open/Closed, Energized/Deenergized) of each of the following:

      i. FWD coil ________ ii. REV coil ________ iii. TD1 coil ________

      iv. TD1 contact ________ v. TD2 coil ________ vi. TD2 contact ________

   c. After the time delay period record the state (Open/Closed, Energized/Deenergized) of each of the following:

      i. FWD coil ________ ii. REV coil ________ iii. TD1 coil ________

      iv. TD1 contact ________ v. TD2 coil ________ vi. TD2 contact ________

   d. Which contact prevent the motor from being started in the reverse direction until the timer has timed out? ________

   e. What is the time of the time-delay period? __________ seconds.

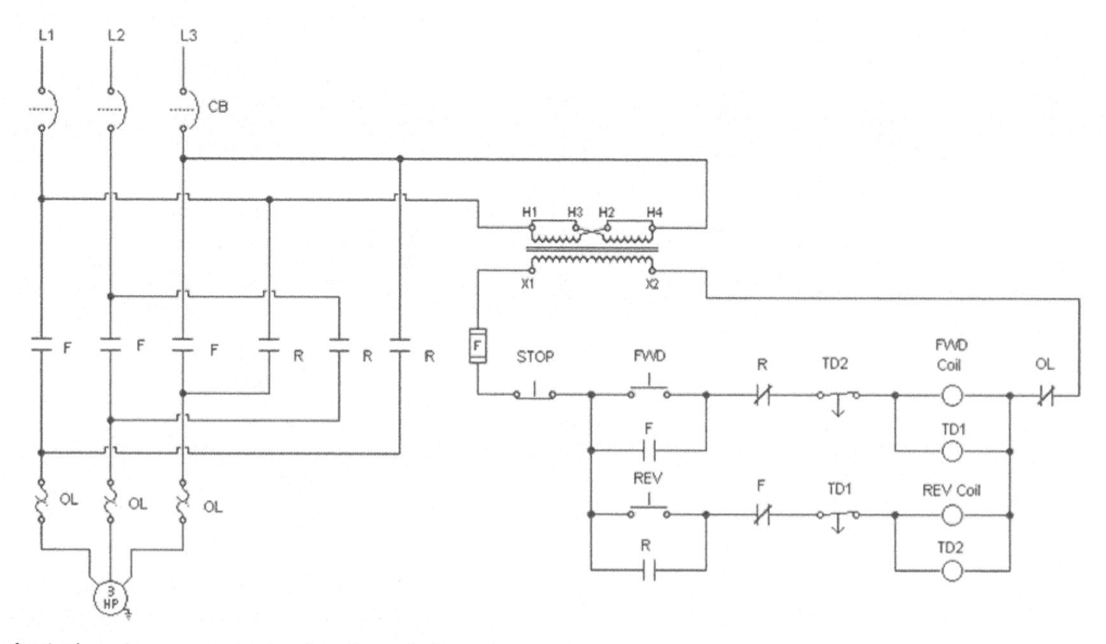

Anti-plugging executed using time-delay relays.

TROUBLESHOOTING ASSIGNMENTS

1. This assignment involves the **troubleshooting** of an **inoperative** two start/stop station motor control circuit. Download the troubleshooting assignment circuit file **CH8 T01** from the **Diagrams-Troubleshooting-NEMA** folder.
   a. Operate the circuit to confirm this problem. With the main breaker closed use the test probe Power function to record the (ON/OFF) power state of the following test points:
      i. Load side of the main breaker (all three lines). ______________
      ii. H1 to H4 of the transformer. ______________
      iii. X1 to X2 of the transformer. ______________
      iv. X2 to both sides of Stop-1 (button not activated). ______________
      v. X2 to both sides of Stop-1 (button activated). ______________
      vi X2 to both sides of Stop-2 (button not activated). ______________
      vii. X2 to both sides of Stop-2 (button activated). ______________
      viii. X2 to both sides of Start-1 (button not activated). ______________
      ix. X2 to both sides of Start-1 (button activated). ______________
   b. Base on your readings, what is the most probable problem? Why?

      ____

      ____

   c. Explain how the test probe Continuity test would confirm this.

      ____

      ____

      (Turn the Main Power Switch off, make the necessary replacement, and operate the circuit to verify your answer.)

2. This assignment involves the **troubleshooting** of an **inoperative** automatic sequential starting of two motors circuit. Download the troubleshooting assignment circuit file **CH8 T02** from the **Diagrams-Troubleshooting-NEMA** folder.

   a. Initially try operating the circuit. What problem, if any, is evident upon start-up?

   ____

   b. With the main breaker still closed use the test probe Power function to record the (ON/OFF) power state of the following test points:

      i. Load side of the main breaker (all three lines). ______________

      ii. H1 to H4 of the transformer. ______________

      iii. X1 to X2 of the transformer. ______________

      iv. X2 to the stop button side of the fuse. __________

      v. Record the initial problem and replace the faulted component.

   ____

   c. After replacement of the component, try operating the circuit again. Why does the same problem occur?

   ____

   ____

   d. Open the main CB. Left-click on the terminal screw to open the parallel circuit between the M1 and TR coils. Use the test probe Continuity function to record the (Yes/No) continuity state of the following control circuit test points:

      i. Across M1 coil. __________

      ii. Across TR coil. __________

      iii. Across M2 coil. __________

   e. According to your readings, what is the most likely problem? ____________
   (Turn the Main Power Switch off and make the two necessary component replacements. Turn the power back on, click the terminal connection back on, and operate the circuit to verify your answer.)

3. This assignment involves the **troubleshooting** of an **inoperative** sequential stopping of two motors control circuit. Download the troubleshooting assignment circuit file **CH8 T03** from the **Diagrams-Troubleshooting-NEMA** folder.
   a. Operate the circuit and observe its operation the following states of its operation:
      i. The ON/OFF state of the motors when the process is initially started.
         Motor-1 ________________________________
         Motor-2 ________________________________
      ii. The state of the motors after continuous operation.
         Motor-1 ________________________________
         Motor-2 ________________________________
      iii. Based on your observations of the circuit operation, describe the most likely problem.
         ____
         ____
         ____
   b. With circuit still operating in the faulted mode, the test probe Continuity function to verify the fault. Describe the results of this Continuity test.
      ____
      ____
      (Turn the Main Power Switch off, make the necessary replacement, and operate the circuit to verify your answer.)

4. This assignment involves the **troubleshooting** of a **faulty operative** primary resistance AC starter circuit. Download the troubleshooting assignment circuit file **CH8 T04** from the **Diagrams-Troubleshooting-NEMA** folder.
   a. Operate the circuit and summarize the faulty condition that exists.

      ____

      ____

      ____

   b. Base on your understanding of how the circuit operates, what can you conclude as most likely being the problem?

      ____

   c. Explain what led you to this concussion.

      ____

      ____

      (Turn the Main Power Switch off, make the necessary replacement, and operate the circuit to verify your answer.)

5. This assignment involves the **troubleshooting** of a **faulty operative wye-delta starter** circuit. Download the troubleshooting assignment circuit file **CH8 T05** from the **Diagrams-Troubleshooting-NEMA** folder.

   a. Operate the circuit and summarize the faulty condition that exists.

      ____

      ____

      ____

      ____

   b. Use the test probe Continuity function to test the following control circuit loads for a short circuit condition. Record the (Yes/No) continuity state of each of the following coils. Keep in mind that a short is indicated by a YES continuity.

      i. Across coil S. __________

      ii. Across coil M1__________

      iii. Across coil M2. __________

      iv. Across coil TR. __________

   c. Use the test probe Continuity function to test the following load circuit contacts for a closed or open condition. Record the (Yes/No) continuity state of each of the following contacts. Keep in mind that a closed circuit is indicated by a YES continuity.

      i. Across each M1 contact. __________

      ii. Across each M2 contact. __________

      iii. Across each S contact. __________

   d. According to your readings, what is the most likely problem?

      ____

      (Turn the Main Power Switch off, make the necessary replacement, and operate the circuit to verify your answer.)

6. This assignment involves the **troubleshooting** of a **faulty operative reversing starter with auxiliary contactor interlocking**. Download the troubleshooting assignment circuit file **CH8 T06** from the **Diagrams-Troubleshooting-NEMA** folder.

   a. Operate the circuit and summarize the faulty condition that exists.

   ____

   ____

   ____

   b. From your understanding of the circuit's normal operation, what specific functional part of the circuit should be suspect?

   ____

   c. Use the test probe to test the part of the circuit that is suspect, and record faulted operating condition you observe.

   ____

   ____

   d. What specific test measurement(s) led you to this conclusion?

   ____

   ____

   (Turn the Main Power Switch off, make the necessary replacement, and operate the circuit to verify your answer.)

7. This assignment involves the **troubleshooting** of a **faulty operative** reversing starter with both electrical auxiliary contactor interlocking and pushbutton interlocking. Download the troubleshooting assignment circuit file **CH8 T07** from the **Diagrams-Troubleshooting-NEMA** folder.
   a. Operate the circuit and summarize the faulty condition that exists.

   ____

   ____

   b. From your understanding of the circuit's normal operation, what two parts of the circuit should be suspect?

   ____

   c. Open the Main CB and use the test probe Continuity function to test the control circuit. Record the (Yes/No) continuity state of each of the following test points:
      i. X2 to the line side of R pushbutton contact. ____________
      ii. X2 to the load side of R pushbutton contact. ____________
      iii. X2 to the line side of FWD. ____________
      iv. X2 to the load side of FWD (button not activated). ____________
      v. X2 to the load side of FWD (button activated). ____________
      vi. X2 to the line side of R aux contact (button activated). ____________
      vii. X2 to the coil side of R aux contact (button activated). ____________
   d. Close the main breaker and with the FWD pushbutton of the control circuit not activated use the test probe Power function to record the (ON/OFF) power state of the following test points:
      i. Line side of the F power contacts (all three lines). ____________
      ii. Load side of the F power contacts (all three lines). ____________
   e. With the main CB still closed momentarily activate the FWD pushbutton and use the test probe Power function to record the (ON/OFF) power state of the following test points:
      i. Line side of the F power contacts (all three lines). ____________
      ii. Load side of the F power contacts (all three lines). ____________
   f. According to your readings, what is the most likely problem?

   ____

   g. What specific test measurement(s), led you to this conclusion?

   ____

   ____

   ____

   (Turn the Main Power Switch off, make the necessary replacement, and operate the circuit to verify your answer.)

8. This assignment involves the **troubleshooting** of a **faulty operative reversing** starter with limit switches connected to limit travel. Download the troubleshooting assignment circuit file **CH8 T08** from the **Diagrams-Troubleshooting-NEMA** folder.

   a. Operate the circuit and summarize the faulty condition that exists.

      ____

      ____

   b. With the main breaker closed use the test probe Power function to record the (ON/OFF) power state of the following test points:

      i. X2 to the line side of F pushbutton interlock contact. __________

      ii. X2 to the load side of F pushbutton interlock contact. _________

      iii. X2 to the load side of the REV button (REV button activated). ________

      iv. X2 to the line side of the limit switch (REV button activated).________

      v. X2 to the load side of the limit switch (REV button activated). ________

      vi. Across the REV coil (REV button activated). _________

   c. According to your readings, what is the most likely problem?

      ____

   d. Explain what reasoning led you to this conclusion.

      ____

      ____

      (Turn the Main Power Switch off, make the necessary replacement, and operate the circuit to verify your answer.)

9. This assignment involves the **troubleshooting** of a **faulty operative** reciprocating machine process. Download the troubleshooting assignment circuit file **CH8 T09** from the **Diagrams-Troubleshooting-NEMA** folder.
   a. Operate the circuit and summarize the faulty condition that exists.

      ____

      ____

      ____

   b. It has been decided to make a continuity test of the most suspected components in the control circuit, as the first attempt to solving the problem. With the main breaker open, use the test probe Continuity function to record the operating continuity (ON/OFF) of the following test points:
      i. Across LS2 normally closed (not activated). __________
      ii. Across LS2 normally closed (activated). __________
      iii. Across the FWD button (not activated). __________
      iv. Across the FWD button (activated). __________
      v. Across the FWD coil. __________
      vi. Across LS1 normally open (not activated). __________
      vii. Across LS1 normally open (activated). __________
      viii. Across LS1 normally closed (not activated). __________
      ix. Across LS1 normally closed (activated). __________
      x. Across the REV button (not activated). __________
      xi. Across the REV button (activated). __________
      xii. Across the REV coil. __________
      xiii. Across LS2 normally open (not activated). __________
      xiv. Across LS2 normally open (activated). __________
   c. According to your readings, what is the most likely problem?

      ____

   d. What specific reading led you to this conclusion.

      ____

      (Turn the Main Power Switch off, make the necessary replacement, and operate the circuit to verify your answer.)

10. This assignment involves the **troubleshooting** of a **faulty operative** pushbutton jog circuit. Download the troubleshooting assignment circuit file **CH8 T10** from the **Diagrams-Troubleshooting-NEMA** folder.

    a. Operate the circuit and summarize the faulty condition that exists.

    ____

    ____

    b. With the circuit operating in the run mode use the test probe Power function to record the (ON/OFF) power state of the following test points:

        i. X2 to the line side of the M axillary contact. __________

        ii. X2 to the jog side of the of the M axillary contact. __________

        iii. X2 to the line side of the jog NC contact. __________

        iv. X2 to the coil side of the jog NC contact. __________

        v. X2 to the line side of the jog NO contact. __________

        vi. X2 to the coil side of the jog NO contact. __________

    c. According to your readings, what is the most likely problem?

    ____

    ____

    d. How would the test probe Continuity function be used to confirm this problem?

    ____

    (Turn the Main Power Switch off, make the necessary replacement, and operate the circuit to verify your answer.)

11. This assignment involves the **troubleshooting** of a **faulty operative** control relay jog circuit. Download the troubleshooting assignment circuit file **CH8 T11** from the **Diagrams-Troubleshooting-NEMA** folder.

   a. Operate the circuit and summarize the faulty condition that exists.

   ____

   ____

   b. With the main breaker open, use the test probe Continuity function to record the operating continuity (ON/OFF) of the following components:

      i. Across the Stop button (not activated). __________
      ii. Across the Stop button (activated). __________
      iii. Across the Start button (not activated). __________
      iv. Across the Start button (activated). __________
      v. Across the CR coil. __________
      vi. Across CR-1 normally open contact. __________
      vii. Across starter coil M. __________
      viii. Across the NO Jog button (not activated). __________
      ix. Across the NO Jog button (activated). __________
      x. Across CR-2 normally open contact. __________
      xi. Across M auxiliary contact. __________

   c. According to your readings, what is the most likely problem?

   ____

   d. Explain in what way the connection of the circuit was altered by the fault.

   ____

   ____

   (Turn the Main Power Switch off, make the necessary replacement, and operate the circuit to verify your answer.)

12. This assignment involves the **troubleshooting** of a **faulty operative** start/stop/selector jog circuit. Download the troubleshooting assignment circuit file **CH8 T12** from the **Diagrams-Troubleshooting-NEMA** folder.

    a. Operate the motor in the pulsating short period jogging mode and record, what problem, if any, is evident.

    ____

    ____

    b. Operate the motor in the run mode for a longer period of time and record what problem, if any, is evident.

    ____

    ____

    c. According to your observations, what is the most likely problem?

    ____

    ____

    (Turn the Main Power Switch off, make the necessary replacement, and operate the circuit to verify your answer.)

13. This assignment involves the **troubleshooting** of a **faulty operative** antiplugging protection circuit. Download the troubleshooting assignment circuit file **CH8 T13** from the **Diagrams-Troubleshooting-NEMA** folder.

    a. Operate the motor in the Forward and Reverse directions. What operating fault condition exists?

    ____

    ____

    b. With the main breaker open, use the test probe Continuity function to record the operating continuity (ON/OFF) of the following components:

    i. Across the Speed SW-R contact (not activated). __________

    ii. Across the Speed SW-R contact (activated). __________

    iii. Across the FWD button (not activated). __________

    iv. Across the FWD button (activated). __________

    v. Across the R normally closed auxiliary contact. __________

    vi. Across the FWD coil. __________

    vii. Across the F normally open auxiliary contact. __________

    viii. Across the Speed SW-F contact (not activated). __________

    ix. Across the Speed SW-F contact (activated). __________

    x. Across the REV button (not activated). __________

    xi. Across the REV button (activated). __________

    xii. Across the F normally closed auxiliary contact. __________

    xiii. Across the REV coil. __________

    xiv. Across the R normally open auxiliary contact. __________

    c. According to your readings, what is the most likely problem? Why?

    ____

    ____

    ____

    (Turn the Main Power Switch off, make the necessary replacement, and operate the circuit to verify your answer.)

14. This assignment involves the **troubleshooting** of an **inoperative** dynamic braking circuit for a DC motor. Download the troubleshooting assignment circuit file **CH8 T14** from the **Diagrams-Troubleshooting-NEMA** folder.
    a. Closed the main breaker and attempt to start the motor by momentarily closing the Start button. With circuit still in this state, use the test probe set to power record the (ON/OFF) power state of the following control circuit test points:
        i. L2 side of the OL contact to line side of the Stop button. __________
        ii. L2 side of the OL contact to load side of the Stop button. __________
        iii. L2 side of the OL contact to line side of the Start button. __________
        iv. L2 side of the OL contact to load side of the Start button. __________
        v. L2 side of the OL contact to line side of M axillary contact. __________
        vi. L2 side of the OL contact to load side of M axillary contact. __________
        vii. Across the Starter Coil. __________
    b. With the circuit still in this mode use the test probe Power function to record the (ON/OFF) power state of the following test points of the power circuit:
        i. Line 2 to the line side of the normally open M contact. __________
        ii. Line 2 to load side of the normally open M contact. __________
        iii. Line 2 to line side of the OL heater element. __________
        iv. Line 2 to load side of the OL heater element. __________
        v. Across the motor ARM. __________
        vi. Line 2 to line side of the normally closed M contact. __________
        vii. Line 2 to load side of the normally closed M contact. __________
        viii. Across the Braking Resistor. __________
    c. According to your readings, what is the most likely problem? Why?

        ____

        ____

        ____

        (Turn the Main Power Switch off, make the necessary replacement, and operate the circuit to verify your answer.)

15. This assignment involves the **troubleshooting** of a **faulty operative** AC motor DC injection braking circuit. Download the troubleshooting assignment circuit file **CH8 T15** from the **Diagrams-Troubleshooting-NEMA** folder.
    a. Operate the motor in the start and stop modes and make note of what operating fault condition exists?

    ____

    b. With the main breaker open, use the test probe Continuity function to record the (Yes/No) continuity state of the following control circuit test points:
        i. Across the starter coil. __________
        ii. Across the TR coil. __________
        iii. Across the brake coil. __________
        iv. Across the Start normally closed contact (not activated). __________
        v. Across the Start normally closed contact (activated). __________
        vi. Across the normally open TR contact. __________
        vii. Across the normally closed M2 contact. ____________ ____________
    c. According to your readings, what is the most likely problem?

    ____

    (Turn the Main Power Switch off, make the necessary replacement, and operate the circuit to verify your answer.)

16. This assignment involves the **troubleshooting** of a **faulty operative** circuit consisting of three motors that must be turned on in order. Download the troubleshooting assignment circuit file **CH8 T16** from the **Diagrams-Troubleshooting-NEMA** folder. The maintenance report from the operator states that the circuit works normally when initially started in the 1-2-3 operating sequence. However, upon momentarily activating PBS1 stop button, Motors M1 and M2 stop operating but Motor M3 continues to run. Operate the circuit to confirm the reported fault.
    a. Reset the simulation by toggling the main breaker on and off. Momentarily activate the PBS1 Start button and use the test probe Power function to record the (ON/OFF) power state of the following test points of the M1 control circuit:
        i. L2 to both sides of the PBS1 Stop button. __________
        ii. L2 to both sides of the PBS1 Start button. __________
        iii. Across the M1 coil. __________
    b. Momentarily activate the PBS2 Start button. Continue your testing and record the (ON/OFF) power state of the following test points of the M2 control circuit:
        i. L2 to both sides of the M1 contact. __________
        ii. L2 to both sides of the PBS2 Stop button. __________
        iii. L2 to both sides of the PBS2 Start button. __________
        iv. Across the M2 coil. __________
    c. Momentarily activate the PBS3 Start button. Continue your testing and record the (ON/OFF) power state of the following test points of the M3 control circuit:
        i. L2 to both sides of the M2 contact. __________
        ii. L2 to both sides of the PBS3 Stop button. __________
        iii. L2 to both sides of the PBS3 Start button. __________
        iv. Across the M3 coil. __________
    d. Momentarily activate the PBS1 stop button. Continue your testing by recording the (ON/OFF) power state of the following test points of the M3 control circuit:
        i. L2 to both sides of the M2 contact. __________
        ii. L2 to both sides of the PBS3 Stop button. __________
        iii. L2 to both sides of the PBS3 Start button. __________
        iv. Across the M3 coil. __________
    e. According to your readings, what is the most likely problem? Why?

    ____

    ____

    (Turn the Main Power Switch off, make the necessary replacement, and operate the circuit to verify your answer.)

17. This assignment involves the **troubleshooting** of a **faulty operative** circuit consisting of two motors interlocked so that the two motors be turned off in order. Download the troubleshooting assignment circuit file **CH8 T17** from the **Diagrams-Troubleshooting-NEMA** folder.

    a. Operate the process in the start and stop modes and make note of what happens.

    ____

    ____

    b. What type of fault should be suspected?

    ____

    c. From reading the circuit, what is the most likely component failure?

    ____

    d. Explain why you concluded this.

    ____

    ____

    e. What test probe test would you do to confirm this?

    ____

    f. What indication should you get?

    ____

    (Turn the Main Power Switch off, make the necessary replacement, and operate the circuit to verify your answer.)

18. This assignment involves the **troubleshooting** of a **faulty operative** Jog/Run selector switch reversing starter. Download the troubleshooting assignment circuit file **CH8 T18** from the **Diagrams-Troubleshooting-NEMA** folder.

    a. Operate the motor in the forward/reverse, Jog/Run modes and make note of what operating fault condition exists?

       ____

    b. With the main breaker open, use the test probe Continuity function to record the (Yes/No) continuity state of the following operating conditions of control circuit test points:

       i. Across the FWD button (nonactivated). __________
       ii. Across the FWD button (activated). __________
       iii. Across the REV button (REV **button head** nonactivated). __________
       iv. Across the REV button (REV **button head** activated). __________
       v. Across the REV normally closed aux contact. __________
       vi. Selector switch common to Run position (Run selected). __________
       vii. Selector switch common to Jog position (Jog selected). __________
       viii. Across the F normally open aux contact. __________
       ix. Across the REV button (nonactivated). __________
       x. Across the REV button (activated). __________
       xi. Across the FWD button (FWD **button head** nonactivated). __________
       xii. Across the FWD button (FWD **button head** activated). __________
       xiii. Across the FWD normally closed aux contact. __________

    c. According to your readings, what is the most likely problem? Why?

       ____

       ____

       ____

       (Turn the Main Power Switch off, make the necessary replacement, and operate the circuit to verify your answer.)

19. This assignment involves the **troubleshooting** of a **faulty operative** anti-plugging circuit. Download the troubleshooting assignment circuit file **CH8 T19** from the **Diagrams-Troubleshooting-NEMA** folder.

    a. Operate the circuit and make note of what operating fault condition exists?

    ____

    b. With the main breaker open, use the test probe Continuity function to record the (Yes/No) continuity state of the following control circuit test points:

    i. Across the FWD button (not activated). __________

    ii. Across the FWD button (activated). __________

    iii. Across the R auxiliary contact. __________

    iv. Across the TD2 timer contact. __________

    v. Across the FWD coil. __________

    vi. Across the TD1 coil. __________

    vii. Across the REV button (not activated). __________

    viii. Across the REV button (activated). __________

    ix. Across the F auxiliary contact. __________

    x. Across the TD1 timer contact. __________

    xi. Across the REV coil. __________

    xii. Across the TD1 coil. __________

    c. According to your readings, what is the most likely problem? Why?

    ____

    ____

    (Turn the Main Power Switch off, make the necessary replacement, and operate the circuit to verify your answer.)